The Differentiated Countryside

In the wake of BSE, the threat to ban fox hunting, and Foot and Mouth disease, the English countryside appears to be in turmoil. Long-standing uses of rural space, such as agriculture, are in crisis. At the same time, urban and ex-urban residents are expressing a whole host of new demands. Not surprisingly, political processes in rural areas are increasingly marked by conflicts between groups, such as farmers, environmentalists, developers and local residents.

The Differentiated Countryside argues that these conflicts emerge because a *national* hierarchy of land uses – which prioritised agricultural requirements in the countryside above all others – has been displaced by a set of *regionalised* approaches. Using an innovative theoretical approach based on 'networks of conventions', the book investigates the 'regionalisation' of the English countryside through a series of case studies. These case studies are based on a set of 'ideal types': 'the preserved' countryside, where environmental pressures are strongly expressed; the 'contested' countryside, where development processes are shaped by disputes between agrarian and environmental interests; and, lastly, the 'paternalistic' countryside, where large landowners continue to oversee patterns of land development. It looks in detail at landowners, traditional and middle-class residents, politicians, planners, farmers, and environmentalists and shows how these groups compete to influence patterns of land development.

The Differentiated Countryside suggests that the complex nature of the new demands being made on the countryside marks the end of the post-war system of national regulation. It argues that the countryside is now increasingly governed by regional policies, formulated and implemented by new regional institutions. It therefore becomes hard to discern a single English countryside; rather we see the emergence of multiple country*sides*, places where diverse modes of identity are expressed and where differing forms of development take place. Such diversity, it is argued, now lies at the heart of rural England.

Jonathan Murdoch is Professor of Environmental Planning in the Department of City and Regional Planning, Cardiff University. **Philip Lowe** is the Duke of Northumberland Professor of Rural Economy in the Department of Agricultural Economics and Food Marketing at the University of Newcastle. **Neil Ward** is Professor of Human Geography in the School of Geography at the University of Leeds. **Terry Marsden** is Professor and Head of Department in the Department of City and Regional Planning, Cardiff University.

Routledge Studies in Human Geography

This series provides a forum for innovative, vibrant, and critical debate within Human Geography. Titles will reflect the wealth of research which is taking place in this diverse and ever-expanding field.

Contributions will be drawn from the main sub-disciplines and from innovative areas of work which have no particular sub-disciplinary allegiances.

1. **A Geography of Islands**
Small island insularity
Stephen A. Royle

2. **Citizenships, Contingency and the Countryside**
Rights, culture, land and the environment
Gavin Parker

3. **The Differentiated Countryside**
Jonathan Murdoch, Philip Lowe, Neil Ward and Terry Marsden

4. **The Human Geography of East Central Europe**
David Turnock

5. **Imagined Regional Communities**
Integration and sovereignty in the global south
James D. Sidaway

6. **Mapping Modernities**
Geographies of central and eastern Europe, 1920–2000
Alan Dingsdale

7. **Rural Poverty**
Restricted lives and social exclusion
Paul Milbourne

8. **Poverty and the Third Way**
Colin C. Williams and Jan Windebank

The Differentiated Countryside

Jonathan Murdoch, Philip Lowe,
Neil Ward and Terry Marsden

LONDON AND NEW YORK

First published 2003
by Routledge
2 Park Square, Milton Park, Abingdon, Oxfordshire OX14 4RN

Simultaneously published in the USA and Canada
by Routledge
711 Third Avenue, New York, NY 10017

Routledge is an imprint of the Taylor & Francis Group

First issued in paperback 2011

Typeset in Galliard by
BOOK NOW Ltd

British Library Cataloguing in Publication Data
A catalogue record for this book is available from the British Library

Library of Congress Cataloging in Publication Data
A catalog record for this book has been requested

ISBN13: 978-1-857-28895-7 (hbk)
ISBN13: 978-0-415-51615-0 (pbk)

Contents

Acknowledgements

This book examines the emergence of rural regions in the English countryside. In developing our analysis of regionalisation processes we draw upon research materials gathered during a number of differing research projects. The information from these projects is presented in Chapters 4, 5 and 6. In Chapter 4 we utilise findings from two projects funded by the Economic and Social Research Council (ESRC). The first – *Exclusive space: networks of participation and forward planning* – was conducted during 1994–5 as part of the ESRC's Local Governance Programme. It examined the range of networks participating in the review of Buckinghamshire's Structure Plan. We interviewed the main participants in the review (including planners, politicians, residents' groups, environmental groups, developers and so on) and attended many of the key review meetings. The second project – *Planning as metaphor: mediating aspirations for community and environment* – was conducted during 1996–7. Here we looked at planning in the district of Aylesbury Vale in Buckinghamshire and paid particular attention to participation at the village level. Again we interviewed the main participants and conducted participant observation exercises at key meetings. We would like to thank Simone Abram who worked on both these projects and was responsible not only for collecting and analysing research data but also for developing some of the ideas expressed in Chapter 4.

In Chapters 5 and 6 we present material collected under another ESRC project – *Environmental action and the policy process: the case of the CPRE* – conducted during 1999–2000 as part of the ESRC's Democracy and Participation Programme. This project looked in detail at the Council for the Protection of Rural England (CPRE). We examined the group's activities at the central, regional and county levels of organisation. We studied three branches in detail – Hertfordshire (not referred to in the follow chapters); Devon (Chapter 5) and Northumberland (Chapter 6) – and interviewed CPRE activists in various organisational positions, as well as politicians, civil servants, planners and local councillors who deal with the group on a regular basis. We would like to thank the researcher on this project – Andrew Norton – for his valuable work in gathering research materials and in helping to develop the analysis outlined in Chapters 5 and 6. In Chapter 6 we also draw upon data collected as part of unfunded research examining landownership structures in the North East of England. This research involved interviews with the largest landowners in Northumberland and others who are engaged in the land management process, such as estate managers and land agents. We are grateful to our respondents, especially those who generously permitted their names to be used in the chapter.

These various projects involved semi-structured interviews, participant observation, focus groups and secondary data analysis. We have woven interview and focus group material into the three case study chapters and we include quotations taken from the fieldwork. Where a quotation is provided and no reference is given it can be assumed that the material was gathered as part of the research work mentioned above. In general, we have maintained the anonymity of those we have quoted, though we have sought to make clear their official positions. In using this material we acknowledge our gratitude to all our respondents in Buckinghamshire, Devon and Northumberland for giving generously of their time and for sharing their stocks of 'local' knowledge with the research team.

We are especially grateful to the ESRC for funding not only three of the four research projects mentioned above but also for sponsoring the Countryside Change Programme which ran between 1988 and 1993. This Programme, which examined land development processes both across England as a whole and within three counties (Buckinghamshire, Cumbria and Devon), allowed the authors to gather the research materials that are included in the first two books in the *Restructuring Rural Areas* series – *Constructing the countryside* (1993) and *Reconstituting rurality* (1994). We would like to extend our gratitude to Andrew Flynn, Julie Grove-Hills and Richard Munton who were part of the original Countryside Change research team and who helped to develop many of the themes that ran through the Programme and which recur in this, the last book in the series.

Introduction

Two main narratives act to shape perceptions of the British or, more accurately, the English countryside. The first narrative, which we might term *pastoralism*, finds its roots in the Romantic movement of the eighteenth and nineteenth centuries. Romanticism emerged in response to the profound changes in economy, society and environment wrought by the Industrial Revolution. In responding to these changes, writers such as Wordsworth, Ruskin and Tennyson extolled the virtues of 'unsullied' nature as an antidote to the corrupted industrial city. Nature was to be found in areas where the degenerative impacts of urban and industrial society were largely absent. Thus, the countryside came to be valued as a zone that lies 'beyond' industrialism (R. Williams, 1973).

As Jan Marsh (1982) demonstrates, the pastoral impulse, although rooted in Romanticism, acquired further validation following the agricultural depression in the latter part of the nineteenth century. The visible decline of rural areas at this time provoked a rush of nostalgia for rural life:

> with the traditional countryside of England apparently disappearing forever, pastoral attitudes were re-asserted with intensity. The city was seen as physically and morally corrupting. . . . Health and happiness were only to be found in the country, in rural life and rural occupations.
>
> (ibid.: 4)

However, the re-assertion of Romantic or pastoral ideas was now undertaken not by an influential elite of poets, writers and other intellectuals but by a broad-based, popular movement associated with the expanding middle class. Moreover, the pastoral impulse manifested itself in rather mundane ways as middle-class households moved away from city centres into the growing suburbs and surrounding rural areas. As Marsh puts it:

> The feeling for the countryside which led many people to dream of living there combined the residual Romantic impulse that had first made the wild places of Europe alluring but was now petering out in a vague love of rusticity, with growing anti-industrialism. Emotional value accrued to the

> farmland that was no longer England's economic base and the countryside became a symbol of escape from the dominant values of capitalism.
>
> (ibid.: 27–8)

'Escaping' to the countryside has remained a dominant theme in English culture. And with the inexorable growth of the middle class during the twentieth century, more and more members of society have come under its spell so that now a majority of households appear to hold the aspiration for a rural lifestyle.

Widespread adherance to pastoralist thinking shows that though the contemporary middle class is an urban class, its members are motivated by strong elements of *anti-urbanism*. This anti-urbanism has two main components. First, the middle class expresses a form of *communitarianism*, one that extols the virtues of the small-scale, traditional communities that are thought to inhabit the typical English village. Where the city is seen as exclusionary and alienating, the village is seen as inclusive and close-knit. The 'true' community resides in the village for only here is the scale of social life tailored to the needs of the individual. Second, the middle class holds a *pre-industrial* view of the countryside. This derives from the Romantic perception that the countryside should remain free from the industrial and urban processes that otherwise shape modern society. Thus, middle-class conceptions of the countryside evolve from older, even aristocratic, notions of the rural aesthetic and the value of rural nature. By escaping the city for the countryside the middle class is somehow escaping the industrial era in order to embrace an older, more traditional way of life (Lowenthal, 1985).

And yet, as the middle class escapes the city and embraces those areas that enshrine its anti-urban sensibilities, it encounters the traditional rural residents that have long lived in such places. Thus, new middle-class residents of the countryside live in close proximity to landowners, farmers, labourers and other traditional rural dwellers. While new and harmonious relationships between these groups can be established, there is also great potential for conflict as each pursues divergent interests. As we shall see in the following chapters, patterns of consensus and contestation between new rural residents and long-standing rural dwellers run through development processes in the contemporary countryside.

The second narrative that has shaped understandings of the countryside is *modernism*. Those characteristics of rural areas that have been so celebrated by the Romantics and pastoralists – tradition, proximity to nature, timelessness, and so on – have been seen by others as signs of backwardness. Thus, depopulation of the countryside during the nineteenth century was not simply the result of the economic forces unleashed by the Industrial Revolution; it stemmed from a widely held view that life would be better in the cities, especially for those working-class households that could expect little more than grinding poverty from life in the countryside. As Marsh again puts it: 'in general, the urban worker looked on his rural counterpart with contempt and pity, as a bumpkin fellow with mud on his boots and straw in his brain' (ibid.: 6). The move to the city on the part of many was therefore a move away from rural backwardness and into the dynamic powerhouses of the modern world.

In opposition to the pastoralists, the modernisers felt that rural areas should be incorporated into industrial society so that the many material benefits of this society could be made available to rural dwellers also. This too has been a constant refrain throughout the nineteenth and twentieth centuries and it has taken a variety of forms. It is evident, for instance, in the belief that households in rural areas are entitled to the same education, health and welfare services as households in urban areas. And indeed, services have improved immeasurably in recent times, with the countryside effectively being encompassed within a national welfare state. The modernising impulse has likewise been extended to transport provision, with first railways and then roads serving to bind rural areas into broader patterns of economic and social development.

Since the Second World War the modernisation discourse has also applied to the rural economy, in particular, to agriculture. For strategic reasons, the state has taken upon itself the task of ensuring that principles of industrial efficiency and productivity become central to agricultural production and it has built up an extensive and elaborate system of intervention in the industry. As a consequence, the agricultural sector has been profoundly re-structured, with successive waves of technical innovation leading to a transformation in working practices, farm structures and rural environments. State agricultural policy has been accompanied by a state planning policy which has also sought to administer rural areas in line with the modernist impulse.

The process of modernisation has integrated rural areas more fully into the capitalist economy. Thus, the countryside has increasingly become subject to processes of rationalisation more generally associated with industrial and urban zones. Even agriculture, an industry that has many traditional and pre-modern characteristics (the family unit, small-scale production units, proximity to nature, etc.), has been re-made by commercial and scientific systems of knowledge and practice. The industry has therefore increasingly changed in line with broad shifts in national and international economic trends.

Pastoralism and modernism provide two contrasting perspectives on the countryside yet both have been influential in shaping not only how we see the countryside but how we act towards it. Each perspective prioritises certain values and actions: on the one hand, we might aim to maintain the countryside as an exclusive and preserved space, one that should be free from industrialism's corrupting influence; on the other hand, the countryside should be brought inside modern patterns of development so that rural resources can make a full contribution to national well-being. Put in these terms, it is clear that in vital respects these two perspectives are in conflict with one another. The Romantics and the pastoralists see the countryside as an area to be protected from modernity, as a place that should be maintained as a pre-modern space; the modernisers see the countryside as inherently backward, as needing reform if it is to achieve the levels of dynamism that are integral to capitalist society.

As we mentioned above, the carriers of these ideas, who often co-exist nowadays within rural communities, can come into conflict with one another. However, we here encounter the first of many ironies in assessing the contemporary countryside, for the strongest protectors of rural areas are often those people who have moved into such areas from urban areas, while the strongest proponents of modernisation are often those who have long resided in rural locations. Thus, the sharp distinctions that would appear to demarcate 'pastoralists' and 'modernisers' begin to blur in practice: pastoralists have only come to live in rural locations because railways and roads have made such areas accessible to them; traditional residents seek to modernise their areas often because they wish to enjoy the benefits of modern society from *within* the rural realm (that is, they have no desire to move to the city to gain such benefits).

The relationship between these two broad perspectives is also re-fashioned by wider changes in the economy, polity and society. On the one hand, the gradual concentration of economic activity in the large towns and cities in the wake of the Industrial Revolution, along with the progressive rationalisation of agriculture, gradually diminished the economic significance of the countryside, thereby allowing it to be re-valued as an aesthetic or environmental space by the pastoralists. On the other hand, the emergence of the national welfare state permitted the extension of economic and social policies into the rural domain thereby incorporating the countryside within the modernising agendas of post-war governments. In both these instances, broad shifts in the surrounding context affected the balance between the two perspectives and their impacts on the countryside.

In our own time, important structural shifts are again taking place: the location of economic activities is changing, with the consequence that rural areas are now more central to new forms of economic activity; the state is re-thinking its means of intervention in economy and society so that resources are re-distributed across rural space; and rural society is being transformed by changes in the class structure, notably the rapid growth of the middle class. Such shifts lead to new economic, political and social alignments in the countryside and these promote re-assessments of the pastoralist and modernist perspectives outlined above.

In this book we seek to identify the main effects of such re-structuring processes in countryside contexts. In particular we assess how structural change is affecting the relationship between rural preservationism (as expressed by pastoralists) and rural developmentalism (as expressed by modernisers). We look in some detail at how new conflicts between the two perspectives are playing themselves out. In making this assessment we pay particular attention to the diverse contexts in which the two narratives are set. We propose that the relationship between the two varies according to the local balance of economic, political and social forces. In some areas we find preservationism to the fore as a strong middle-class constituency re-casts rurality in its own image. In other areas, middle-class residents find their views overridden by established rural perspectives so that a more developmental and modernising pattern of change is evident. In yet other areas some measure of co-existence between the two perspectives is achieved. We investigate these differing contexts in the chapters that follow.

In short, we propose that the contemporary countryside emerges from a rivalry between pastoralists and modernists and the expression of this in contrasting geographical circumstances. We aim to capture various dimensions of the interrelationship between pastoralism and modernism by looking in some detail at processes of *differentiation* in the countryside. The notion of the 'differentiated countryside' explicitly refers to the patterns of geographical diversity that can now be found in rural areas. It also refers to the general modes of development that promote such diversity. In other words, it asserts that we can only make sense of development processes in discrete rural localities and regions by reference to the structures, institutions and organisations that shape the countryside as a whole. Thus, the modes of development that are *internal* to particular rural areas must be linked to the *external* influences upon such areas.

The study we present here takes a 'holistic' approach and considers the interaction between economic processes, political regulation, social structures and environmental change. However, given that the scale of this enterprise is quite vast, we focus our attention on the way these elements interact within land development processes. Land development, as we argue in the first chapter, provides a useful vantage point from which to assess broader patterns of change. While this vantage point does not encompass all aspects of rural life, it does allow us to focus upon the relationships between economic, political and social phenomena and to consider their combined impacts upon the rural environment.

In the main, we examine the development process as a *social* process, one which contains diverse social forces and world views. In contrasting regional settings we seek to describe the dominant formations of actors that determine the social and material shape of rural space. We show how economic, political and social networks come together in order to affect rural land development and we consider the values that bind the involved actors into the networks. We study examples of what Deakin (2002: 60) calls 'spontaneous co-operation', that is, economic, political and social forms in which actors align their own personal interests with conceptions of the 'common good', conceptions that are generated in discrete geographical contexts. The 'common good' is usually articulated in relation to some broader social or spatial entity (the village, the neighbourhood, the region or the country) and these articulations are enacted across various spatial scales.

We present case studies of networking activities in differing rural contexts, contexts that are chosen to show the varied interrelationships between social groups in the countryside. We outline how the interactions between the groups and the networks leads to the consolidation of discrete 'environments of action' in which certain values and modes of behaviour are prioritised. The consolidation of distinct 'action environments' leads to separate trajectories of rural development. In other words, the various networks give rise to regional formations and these formations increasingly undermine national modes of economic and social regulation. It is this broad shift from a national policy space towards a set of regionalised rural spaces that comprises the main theme of our analysis.

This book should therefore be read as an attempt to take stock of the English countryside at the beginning of the twenty-first century. While we do not aim to provide a comprehensive overview of this territory – there are many issues extending from rural poverty to the state of the rural environment that are omitted from the analysis – we do wish to show how contemporary versions of pastoralism come up against contemporary expressions of modernism. In our view, the relationship between these two narratives continues to decisively shape the development of the countryside. However, this relationship varies according to geographical context. As we shall show, differing countryside areas are now valued in a variety of differing ways. These various modes of evaluation give rise to rural regions that are increasingly differentiated from one another. This process of differentiation provides the core theme of the book.

1 A differentiated countryside?

Introduction

The English countryside has never been far from the headlines in recent years. To the casual observer, rural areas and industries seem to lurch from one calamity to the next in a perpetual state of turmoil. A key episode was the Bovine Spongiform Encephalopathy (BSE) crisis which reinforced already widespread perceptions that intensively farmed food could be damaging to health. The identification in 1996 of new variant Creutzfeld Jakobs Disease (CJD) as a fatal human illness sent shock waves through the industry and led to a profound questioning of modern agricultural practices (Hinchcliffe, 2001). This questioning was intensified by the Foot and Mouth crisis of 2001, which turned out to be the most serious outbreak of the disease the world has yet seen. The public bill for Foot and Mouth approached £3 billion and around 6 million animals were slaughtered on 10,000 farms. In addition, a wide range of non-agricultural businesses – from tourism to services – were adversely affected as recreational pursuits in the countryside were halted for several months. The story of the Foot and Mouth crisis is one of the conduct of an agricultural disease leading to a general crisis in the rural economy, prompting demands from rural tourism, recreational and business interests for short-term aid and greater recognition of their significance to the countryside (Bennett *et al.*, 2002).

While animal disorders have caused havoc in both the countryside and the food chain, rural discontent has also been widespread. In recent years there have been successive falls in farm income, such that by 2000 total income from farming was 72 per cent lower than in 1995 (Department of Environment, Food and Rural Affairs [DEFRA], 2002a). In responding to falling prices and rising costs, farmers' groups have blocked roads and picketed supermarkets. These actions culminated in the fuel protests of the summer of 2000 initiated by a loose grouping of farmer activists called 'Farmers for Action'. There have also been large-scale demonstrations in London and regional cities against threats to 'the country way of life', prompted by parliamentary initiatives to ban hunting with hounds. In a context of economic vulnerability, the threat to hunting caused considerable disquiet in rural areas. Opposition to the proposed ban, orchestrated by the Countryside Alliance, emphasised the amount of economic activity and employment associated with hunting, as well as its traditional role in rural society (N. Ward, 1999).

These various protests and concerns have combined to generate a widespread perception that rural areas in England and elsewhere in the UK are in crisis.[1] In the face of repeated expressions of rural discontent, Ministers in Tony Blair's first Labour government became extremely wary of the charge that they had failed to respond adequately to rural problems, and there was an extensive effort in the late 1990s to formulate a new policy framework for the countryside. The Government drew up a White Paper, published in 2000, under the title *Our countryside, our future* (Department of the Environment, Transport and the Regions [DETR] and Ministry of Agriculture Fisheries and Food [MAFF], 2000), which spelled out the broad range of state initiatives that are operational in rural areas. Following the Foot and Mouth crisis of the following year, Tony Blair also established a Policy Commission on the Future of Farming and Food which produced its own report on the future of the countryside entitled *Farming and food: a sustainable future* in early 2002 (see Policy Commission, 2002). The Commission proposed changes in the direction of public support for agriculture and attempted to chart a course out of the crisis.

In reflecting on these events, we can say that rural issues have gained political prominence because some long-standing uses of the countryside – such as agriculture and hunting – are being threatened by contemporary economic and political trends. As a consequence, the hierarchy of activities that has long dominated rural space has been challenged by alternative demands on rural land and other resources. What counts as a legitimate use of land-based resources can no longer be automatically assumed by reference to past practice and consequently activities in a range of sectors have been politicised. In the case of agriculture, environmental, consumer and animal welfare lobbies demand improved standards and changed systems of production. In the case of countryside recreation, argument rages over the acceptability of hunting or the 'right to roam' across privately-owned rural land. In the case of physical planning, local needs for housing or employment are pitched against the demand to protect greenfield land and rural amenity. In the case of rural development, those promoting rural tourism and services challenge the priority of expenditure on farm-based development. These contested issues indicate that many groups now make claims upon rural space, but no single view is able to encompass the whole rural sphere. The outcome is a greater potential for conflict in and around the countryside.

The emergence of more strongly competing claims on the countryside can be linked to broader patterns of economic and social change, notably the decline of national economic and social structures, and the emergence of much more complex networks and flows across territories. Where once rural life was strongly 'framed' by the national context – so that the countryside could be seen as a stable and functional part of a national state that acted to establish land use priorities across rural localities – now any given rural area is constituted by economic, political and social processes operating at transnational, national, regional and local scales. These processes combine in differing ways in differing regions and localities. As a consequence, national land-use hierarchies are now being displaced by more localised arrangements.

The interaction of differing economic, social and cultural processes in the countryside also gives rise to shifting meanings of 'rurality' and the analysis of these meanings has generated a great deal of academic work in recent years. Scholars have looked in detail at the changing nature of the rural economy and rural society (Boyle and Halfacree, 1998; Ilbery, 1998; Phillips, 1998), at the cultural aspects of rural life (Cloke and Little, 1997; Milbourne, 1997), and new socio-cultural practices in the rural domain (Halfacree, 1996; Hill, 2002; Hetherington, 2000). In the light of such work it is now regularly asserted that rurality has splintered into many competing *ruralities*, ruralities that are associated with the various demands on rural space being made by different groups. Furthermore, it is far from clear that these demands have much in common with each other (Marsden, 1998). Thus, it becomes increasingly difficult to identify a single coherent entity called 'the countryside' (see Halfacree, 1993; Hoggart, 1990; Mormont, 1990; Murdoch and Pratt, 1993; 1994; Pratt, 1996). The move away from national hierarchies has therefore opened up scope for multiple forms of rural or country living (Philo, 1992).

Our own contribution to the process of 're-thinking the rural', at least within the context of the *Restructuring Rural Areas Series* of which this book is part, has been to focus on increased conflicts around *land use* (Marsden *et al.*, 1993; Murdoch and Marsden, 1994; Lowe *et al.*, 1997). We have argued that, despite widespread changes in rural economy and society, land remains a special marker of rurality (largely because it is still widely assumed that 'greenness' and 'openness' define the character of the British countryside). Land is also a key resource in the development process and it mediates broader changes in rural economy and society. In our past work we have attempted to show how various actors coalesce around different land uses in the countryside and we have examined land development outcomes in terms of the shifting power relationships between these various actors.

In this book we return to issues of land use and again seek to identify the main economic and social formations that are competing to determine development trajectories in the countryside. In particular, we examine the regional coalitions of actors that operate within land development processes. Moreover, we build upon our previous work by drawing together findings from a number of research projects in order to propose a comparative analytical framework for the study of rural areas. In what follows, this framework is applied to rural localities in England, localities which are set in their regional contexts (in the South East, South West and North East regions). This comparative approach allows us to investigate the changing nature of contemporary rural spaces and to illustrate how rural regions are becoming increasingly differentiated from one another. In other words, we aim to situate the multiple demands being made on the countryside in their various socio-spatial contexts. By specifying the differing contexts that comprise the (English) countryside we seek to identify underlying coherences in contemporary patterns of countryside change.

In proposing the comparative analysis of discrete areas within the differentiated countryside, we take forward insights and ideas generated in our earlier work in this series. In the first book – *Constructing the countryside* (Marsden *et al*., 1993) – we outlined a general theoretical approach and established some analytical themes to guide rural research. In the second book – *Reconstituting rurality* (Murdoch and Marsden, 1994) – we applied our approach to land development processes in the county of Buckinghamshire in South East England. Here we discerned processes of middle-class formation as the most important influence on rural land use. A third book – *Moralising the environment* (Lowe *et al*., 1997) – took farm pollution as its analytical focus and considered how economic and social actors in the countryside compete to characterise and regulate pollution processes, and in so doing re-define the functions of rural space. In short, all three volumes looked at struggles to determine the developmental character of rural areas and indicated that different areas are on distinct trajectories of change.

In this fourth book we explicitly address the contrasting fortunes of rural localities using a comparative approach. Following arguments presented in *Constructing the countryside* concerning the emergence of the so-called 'post-productivist' countryside, we propose that rural areas in the UK are no longer situated within a *national* regime of regulation. Instead, we are witnessing the emergence of *regionalised ruralities*. In other words, rural areas are being integrated into regional formations which are proceeding along their own distinctive trajectories of development. These regionalised rural formations can be seen as representing new forms of coherence in a context of growing economic and social complexity. The analysis of new patterns of regionalisation presented in the following chapters can thus be regarded as an attempt to define some of the main parameters of change that underpin processes of spatial development in rural England and elsewhere.

The Countryside Change Programme: a brief overview

In this first chapter we outline some of the main themes that will guide analysis of the various case study localities and regions presented in later chapters. In elaborating a theoretical framework for the analysis of the differentiated countryside, we take as a starting point our earlier work under the Economic and Social Research Council's (ESRC) 'Countryside Change Programme' and its key conclusions about rurality and development. In this section we briefly review some of the main findings of that earlier project before moving on to outline a revised theoretical framework.

In 1988, the ESRC decided to fund a major research programme to investigate the transformation of rural areas in Britain. The genesis of the ESRC's decision dated back to the mid-1980s and a series of discussions between the ESRC, the then Ministry of Agriculture, Fisheries and Food (MAFF), the then Department of the Environment (DoE), and agencies such as the Countryside Commission. Rural policy-makers were facing new types of challenges and many of the assumptions underpinning the post-war framework for rural and agricultural policy were being undermined. Commentators proclaimed the countryside to be 'at a crossroads' as old priorities around agricultural efficiency and expansion, the role and function of land use planning, and the provision of public services in rural areas came under challenge. New priorities – for homes, for services and amenities, and for greater efforts to preserve valued rural environmental goods – were being asserted. An academic re-assessment of rural space seemed to be required.

One of the main strands of the research that was funded by the ESRC focused upon the social, political and institutional dynamics of contemporary rural change. It adopted a case study approach in analysing the constellation of re-structuring pressures affecting rural localities, and took as its empirical focus the range of land development processes being experienced in rural areas. Land development, it was argued, provides a vantage point from which to view social, political and regulatory change, as well as the economic restructuring processes being played out in rural localities (see Marsden *et al.*, 1993 for a summary of this view). Typical land development processes in rural areas include agricultural intensification, farm diversification, house building, golf-course construction, mineral extraction and waste dumping. It was claimed that not only are these processes altering the social complexion and the economic functions of the countryside, but also that they are contributing to a growing differentiation of spatial areas within rural Britain.

Growing differentiation makes rural areas the loci for new social and political conflicts – notably between economic development and environmental protection. Work conducted under the Countryside Change Programme explored such conflicts by identifying the main actors – developers, politicians, planners, interest groups, local residents, etc. – involved in particular development proposals and then 'followed' these actors as they attempted to shape the final development outcomes. In so doing, it also sought to analyse connections between actors involved in local cases and wider economic, social and political forces. Thus, 'networks' were traced out from local development conflicts to wider institutional and socio-economic structures in order to place local actions in the development process in their respective socio-economic contexts (a mode of analysis that we termed 'action in context'). In this way, rural land use changes were connected to the most significant national and international processes operating in the requisite sector, whether it was housing, minerals, recreation, industrial development, agricultural intensification or environmental regulation.

Two main assumptions lay at the heart of the work conducted under the Countryside Change Programme. First, the active role that local interests play in the development of their localities cannot be ignored. Thus, a top-down causal argument which portrays rural areas as the passive recipients of general movements of capital and labour is inadequate as a basis for explaining the uneven nature of rural development. Local action may be constrained by broader forces, but it cannot be dismissed, especially as the range of new development opportunities is increasing. Second, a concern for local action leads on to an interest in local differentiation in development practices and outcomes. Such differentiation, it was argued, emerges from the interaction between local development actions and broader processes of change. In particular, it was proposed that local differentiation is being encouraged by more flexible systems of production, the growing scale and increasing range of consumption-led demands for rural resources, the deregulatory instincts of central government and the changing nature of agriculture. These trends affect rural areas in differing ways, depending upon the changing mix of local resident populations, the historical pattern of land ownership and capital investment and local environmental endowments. Rural areas are increasingly differentiated from one another.

In *Constructing the countryside*, four main sets of parameters were identified as crucial in shaping the development trajectories of rural localities:

- First, *economic* parameters, including the structure of the local economy and relationships to wider economic forces.
- Second, *social* parameters, including shifting demographic structures and associated patterns of social change.
- Third, *political* parameters such as the organisation of local politics and participation in the development process.
- Fourth, *cultural* parameters, which include dominant attitudes towards property rights, community identity and leisure practices.

The geographically varied interplay of these different parameters creates different 'types' of countryside. For heuristic purposes, four countryside 'types' were identified:

- the *preserved countryside*
- the *contested countryside*
- the *paternalistic countryside*
- the *clientelist countryside*.

The *preserved countryside*, we argued, is evident throughout much of the lowlands of southern England, as well as in attractive and more accessible upland areas. It is characterised by the dominance of pastoral and preservationist attitudes and decision-making processes. Such concerns are expressed mainly by middle-class social groups living in the countryside, employed primarily in the service sector, and often working in nearby urban centres. These groups attempt to impose their views through the planning system on would-be developers. In addition, the existence of this social group in rural areas provides a ready source of demand for new development activities associated with leisure, the service sector and residential property. Thus, the re-constitution of rurality is strongly shaped by articulate consumption interests, who use the local political system to protect their 'positional goods', and the development opportunities that follow as such groups become more numerous in the countryside (Murdoch and Marsden, 1994).

The *contested countryside* refers to areas that lie outside the main commuter zones. Here local agricultural, commercial and development interests may be politically dominant and these interests will tend to favour development for local needs. However, these traditional development interests are increasingly opposed by 'incomers' who may be middle-class workers or retirees attracted to the area because of its residential environment. Thus the development process is marked by increasing conflict between old and new groups, but with no single interest attaining overall dominance (Lowe *et al.*, 1997).

The *paternalistic countryside* refers to areas where large private estates and big farms predominate and where the development process is decisively shaped by established landowners. Many of the large estates and farms may be faced with falling incomes and will therefore be looking for new sources of income. They will seek out diversification opportunities and are likely to be able to implement these relatively unhindered. They tend to take a long-term management view of their properties and adopt a traditional paternalistic role. These areas are likely to be subject to less development pressure than either of the above two types, partly because the middle class is present in much lower numbers (Newby *et al.*, 1978).

Finally, in the *clientelist countryside*, which can be found in remote rural areas, agriculture and its associated political institutions still hold sway. Processes of rural development are determined by state agencies, in part, because farming can only be sustained by state subsidy (such as through Less Favoured Area payments, EU regional funds and welfare transfers). Farmers will be dependent on systems of direct agricultural support and any external investment is likely to be driven by state aid. Thus, state agencies and agricultural interest groups frequently work together in close (corporatist) relationships (Murdoch, 1992). Local politics is dominated by employment concerns and the welfare of the 'community' (Munton, 1995).

We presented these four types in order to indicate that socio-economic processes are 'regionalising the rural': that is, interactions between local, national and international processes are generating new constellations of economic and social relations which are gaining some coherence at the regional level. In this, the last book in the series, we develop further the analysis of regionalised spatial types by presenting comparative work on three localities set within their regional contexts (Figure 1). The areas are, first (in Chapter 4), *Buckinghamshire*, a county that is located close to London in the South East of England and which has been subject to sustained development pressures over the last three decades. Local analysis reveals that Buckinghamshire forms part of an increasingly important category of rural Britain where, despite the development demands created by a buoyant sub-regional economy, social forces in favour of the protection of scarce positional goods are especially strong (Murdoch and Marsden, 1994). A key issue for these forces is how to maintain a sense of rurality in a metropolitan region characterised by rapid growth.

The second case study (presented in Chapter 5) is *Devon* in the South West of England. This area is chosen because it is representative of those rural places that are still dominated by small businesses (e.g. in tourism, manufacturing and farming sectors) but which are highly attractive to in-migrant retirees. Local politics is more fragmented and localistic than in the other study areas and traditionally development has proceeded in line with the aspirations of successful

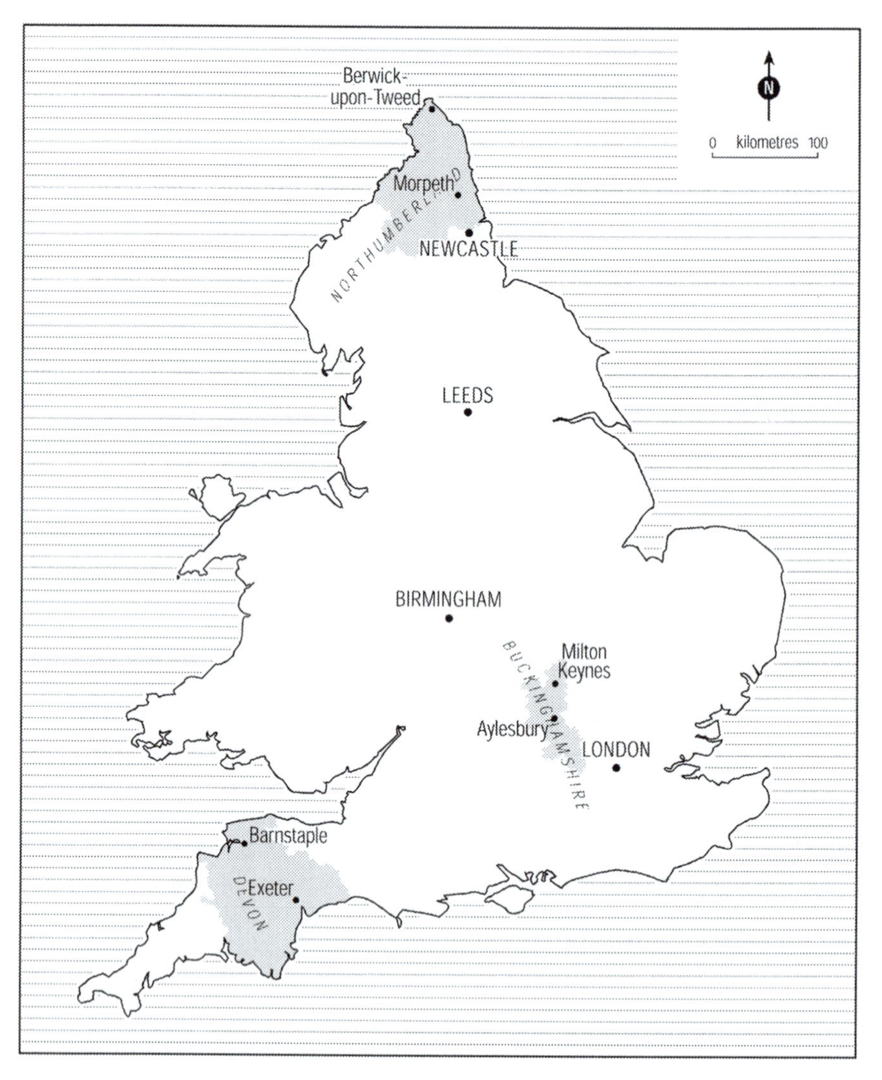

Figure 1 Case study areas.

local entrepreneurs. Development patterns and policies have thus tended to lack coherence or an overall sense of direction. However, the gathering force of local environmentalism, fuelled by the high rate of social and economic change, is beginning to challenge this pragmatic approach to development and it seeks to contain the entrepreneurial independence of existing, and often long-standing, local interests.

The third case study is *Northumberland* in the far north of England (Chapter 6), an area that has a distinct economic history and set of locational influences. The county is struggling to re-build a local economy weakened by the decline of mining and the squeeze on agricultural viability. In doing so, it is significantly disadvantaged by its peripheral location. A key feature of Northumberland is the persistence of large landed estates. Through their extensive landholdings the estates decisively shape patterns of rural development. Moreover, local politicians retain a 'welfarist' approach to rural areas and therefore attempt to orchestrate new forms of economic growth. Taken together, these characteristics create an outlook that favours development over conservation and environmental protection.

In later chapters we compare the three localities and, in so doing, undertake four main tasks. First, we seek to highlight the underlying causes of regionalisation in the British countryside; that is, we attempt to identify theoretically the main features of the differentiated countryside and the processes that are generating these features. Second, we present material that shows how the differing regional types are being consolidated in the English context. As mentioned above, the focus on processes of differentiation requires the comparative analysis of our differing study areas and this task implies a need to scrutinise the utility of the countryside types outlined in our earlier work. Third, we consider the policy processes that are both generating regional outcomes and reflecting such outcomes. In conclusion we assess how the forces of regionalisation are reshaping 'rurality' in contemporary England.

Criticisms of 'Countryside Change'

Before outlining some of the theoretical considerations that will guide our comparative analysis of the study areas, it is worth pausing to consider how the previous volumes in this series have been received. From reviews of the earlier books, and from the subsequent reception of the Countryside Change Programme's research findings in the academic literature, we can identify a number of concerns and reservations about our approach among interested scholars. The first main concern was that the research was pre-occupied with the role of agriculture to the detriment of other important dimensions of rural change. Second, it was claimed that in focusing upon land, property and power, the research did not adequately deal with the 'cultural politics' of rurality. Third, it was implied that the arguments deployed in *Reconstituting rurality* tended to overplay the role and significance of the middle classes in the south of England in comparison with other social groups in other rural areas. Finally, the attempt to move away from overly deterministic structuralist accounts of rural change in order to accommodate a greater role for local social action was seen as lacking theoretical rigour. In this section we briefly consider each of these concerns.

Under the first criticism, some scholars made the point that, in addressing the changing conditions of agriculture, the earlier books had over-emphasised the role and significance of this sector. Hoggart (1994: 91), for example, complained that 'the heavy hand of farming' hangs over *Reconstituting rurality*. As a general view this concern is valid for, in terms of national employment, agriculture is no longer a major sector. By the mid-1990s, the total workforce employed in UK agriculture stood at around 500,000 people (less than 2 per cent of all workers), and contributed an equivalent proportion to national Gross Domestic Product (GDP). Thus, any analysis that equates the countryside or rurality with agriculture alone is likely to be flawed.[2]

However, it is important to recognise that our interest in agriculture stemmed not from the industry's economic significance but from its control over the use of land in rural areas (despite agriculture's declining economic importance it still utilises around 75 per cent of land). Land, we argued, provides a useful prism through which to observe and analyse the changing distribution of property rights (and thus power relations) in the countryside. In other words, land development processes can serve as a 'window' on broader patterns of social and economic change. They can play this analytical role because of the social and political benefits that access to, and control over, property rights continue to bring. Certain forms of land development are restricted to rural space, including agricultural intensification, mineral extraction and golf course construction, while others – such as housing – tend to have distinctive consequences in rural contexts. But with the widening range of demands on rural areas, it is the changing relations between different uses that are of particular significance. *Because* agriculture remains the major rural land use, it comprises a *starting point* for the study of land development.

A second, related argument provoked by the first three books in the series is that they paid insufficient attention to the changing *cultures* of rural areas. Cloke (1993: 366), for example, suggested that in concentrating on land and property, changes such as 'social repositioning according to gender, age, ethnicity, alternativeness and other agencies of marginalisation, or cultural respresentation as of rural idyll, or discursive transformations over the nature of rural lifestyles' were neglected. Certainly, recent years have seen great interest in these issues as a consequence of what some have called a 'cultural turn' among British geographers (see Cloke and Little, 1997; Milbourne, 1997). Propelling this 'turn' has been 'a conceptual fascination with difference' within rural society (Cloke, 1997: 367).

Yet, despite our interest in rural *differentiation*, we do not seek to situate our land development research within this particular genre, partly because we do not accept the assumption that culture is a discrete realm that cannot subsume, or be subsumed within, economy and society.[3] Our approach here is more directly sociological in that we are concerned with the *social relations* that congeal around development and with the *social forces* and *values* that drive the main development trends. In other words, we have chosen to study *differentiation* in the development process as a means of investigating the various social formations that now comprise rural society, specifically through the way these formations come together around property and economic issues in regional contexts. Our overriding interest is in the way social relations interconnect with economic and political relations to shape differentiation in rural development trajectories.

A third response was again made by Hoggart (1997; 1998), who pointed to the dangers of assuming that what happens in the South East of England can be applied to rural areas in England (or Britain) as a whole. Of course, any locality study can be criticised for its geographical specificity, but Hoggart (1998: 383) suggested that 'by focusing on rural zones that have a restricted array of economic bases and social forms, the chances are that a broader understanding of the character of social change in the countryside is diminished'. Yet it was precisely for this reason that we originally devised the typology outlined above. We were concerned that Newby's classic work on paternalism and class relations in Suffolk in the 1970s (Newby, 1977; Newby *et al.*, 1978) was being applied by others to the countryside *as a whole*. The empirical work under the Countryside Change Programme revealed that Newby's model was less applicable in other rural regions. The typology in *Constructing the countryside* was therefore an attempt to conceptualise the ruralities existing in different regional contexts.[4]

Hoggart (1998: 383) called for parallel work in 'those rural zones that still experience economic decline, that are still dominated by small-scale, owner-occupied farms, and that are the preserve of the "traditional" rural middle classes but with few other economic openings that provide work for manual labour'. We fully endorse this call and in this volume we compare three regions selected in line with the countryside types outlined above. Through the comparative approach, we hope to show how the combination of local social formations and general socio-economic processes are resulting in distinct development trajectories for different rural locations.

Finally, a chapter in an edited book (Lowe and Ward, 1997), which presented some of the findings from *Moralising the environment,* prompted a fierce and indignant response from the American Marxist geographer, Richard Walker (Walker, 1997). In general, Walker objected to the use of a network perspective (derived from actor-network theory) to understand the construction of the farm pollution problem in Devon, arguing that what he called the 'methodological extremism' (ibid.: 275) inherent in the approach amounted to a refusal to commit to any political position. The network perspective, he argued, allows us 'to gambol merrily on without any clear notion of social structure at all', without providing insight into whether 'there is a new agrarian regime (or whether any such thing could even exist)' (ibid.: 274).

In response, we agree that the theoretical approach we have adopted in this book series does indeed represent a move away from conventional Marxist accounts and their tendency to see local social change as little more than the outcome of broad structural shifts (see Marsden *et al.*, 1993, especially chapter 6 for a summary of the various arguments). As we will again explain here (in Chapter 3), in order to understand local trajectories of development, and the nature of struggles between different social groups and interests, the role of social agency must be central to any academic analysis. This cannot be achieved only through the analysis of rural 'regimes' and other large-scale structures. Rather, the inquiry must link local action to broader patterns of change in a fashion that gives appropriate weight to each side of the theoretical 'equation'. In what follows we again propose that this linking can be achieved via the network repertoire introduced in the earlier books, although we note the need to link networks to structural contexts.

In sum, the comments and criticisms made in responses to the first three books in the series have been helpful in encouraging us to reflect upon justifications for our overall approach. We would wish, however, to re-assert our faith in the analysis of land-development networks and the social regulation of property rights as an aid to understanding how processes of rural re-structuring come to be played out in contrasting rural localities. Our work thus remains firmly in a post-Marxist tradition as it attempts to engage with new patterns of complexity and fluidity in the construction of rural space. It sees these patterns as bound up with the interpretive and practical activities of economic, political and social actors operating within broad social formations. This theoretical orientation does not mean that we want to be seen as 'philosophising excessively' (as Walker puts it); rather, we take the notion of the 'differentiated countryside' as a starting point in thinking in rather concrete terms about the social, economic and environmental distinctions emerging between rural regions.

Differentiation as 'regionalisation'

While the land development concern that runs through this volume follows from the analyses presented in the first three books, we have developed here an amended theoretical approach in order to incorporate recent theories of 'regionalisation'. Although the term 'region' is often hard to define, this spatial scale appears to be gaining new significance for economists, sociologists, political scientists and geographers (see Amin and Thrift, 1994; Cooke and Morgan, 1998; Keating, 1997; Storper, 1997). For instance, in a review of the economic development literature, Le Gales and Voelzkow say:

> Localities and regions are becoming more important in Europe as units of organisation of social, political and economic life. If nation-states and national regulations carry less weight, it is not surprising to see in the European context the emergence of infra-national territories. The loosening of state constraints both opens opportunities and creates new constraints for cities and regions.
>
> (2001: 18–19)

There are three main dimensions to regionalisation evident in the academic literature. First, economic geographers and others have noticed that new economic structures are congealing at the sub-national level. While these structures do not equate to the same territorial zone in every instance, they tend to cover areas that are more than 'local' but less than 'national'. The term 'region' is therefore used to describe new industrial agglomerations in which firms and associated institutions cluster together in space. Second, recent political and administrative changes – notably a shift to 'multi-level governance' structures – appear to highlight the significance of the regional tier. In the UK such a shift has been evident in devolution to Scotland, Wales and Northern Ireland and in the emergence of Regional Development Agencies (RDAs) and Regional Chambers in England. In the process, the region has gained the potential to become a more autonomous policy-making arena.[5] The emergence of such autonomy has, in turn, fostered an enhanced consciousness of regional contexts among political actors thereby leading to demands for further devolution, notably in the form of elected regional assemblies. Third, social relations have become 'regionalised' as people become more aware of regions as economic and political entities and orchestrate their lives across regional spaces.

Because 'regionalisation' bears a striking resemblance to 'differentiation', we consider in this section how the regionalisation process is occurring at the present time and speculate on how rural areas are being affected. We briefly review the main forms of regional development so as to provide a broad context in which to place our study of the differentiated countryside. We also draw out of this literature two theoretical tools – *networks* and *conventions* – that we intend to utilise in the comparative analysis of regionalised countrysides. As we indicate in the following section, these conceptual tools can help us to identify key points of comparison between our case study areas and will assist us in accounting for patterns of regionalisation in rural England.

Economic regionalisation

Superficially at least, the advent of global and transnational economic forms might be thought to override the economic distinctions that mark places off from one another. However, as Amin and Thrift remark,

> global economy and global society continue to be constructed in and through territorially bound communities . . . metaphors such as 'global village' and 'one world' are complicated, perhaps even contradicted, by the presence of villages, towns, districts, cities, and regions which continue to tell their own stories of economic development and cultural or political distinctiveness.
>
> (1994: 5)

Globalisation, Amin and Thrift (ibid.: 6) conclude, 'does not imply a sameness between places, but a continuation of the significance of territorial diversity and difference' (see also Short, 2001).

According to some accounts, differentiation in the fortunes of territorial areas in the context of an increasingly globalised economy stems from the emergence of a 'post-Fordist' regime of flexible accumulation in which high levels of economic growth derive from clusters of small- and medium-sized firms strongly linked to regional institutions (Amin, 1999). These economic clusters can achieve rapid growth because they more easily facilitate learning and innovation than hierarchical or atomised production structures. Economic and non-economic institutions are bound together by a host of linkages, leading ultimately to the creation of robust industrial agglomerations in which flexibility and specialisation allow easy adjustment to changing economic circumstances. And, as firms cluster more densely together through rich arrays of network linkages, so the economic region comes into view.

In a recent review of the regionalisation process, MacLeod (2001: 809) identifies differing emphases in the literature. He shows that one strand of work has focussed upon the role of urban cores (in the context of a global economy) in orchestrating regional arrangements. In this view, it is the ever-extending spatial geometry of the city that underpins the new industrial structures. However, another analytical strand looks at how economic institutions become embedded in social, political and cultural structures at the regional scale. Scholars working in this vein argue that non-economic institutions serve to generate 'networks, norms, conventions, trust-based (often face-to-face) interactions and horizontal relations of reciprocity' and that these 'enhance the benefits of investments in physical and human labour' in given regional locations (ibid: 808). Powell and Smith-Doerr provide a flavour of the approach when they write that:

> The more distinctive each firm is the more it depends on the success of other firms' products to complement its own. Repetitive contracting, embedded in local social relationships, cemented by kinship, religion and politics, encourages reciprocity. Monitoring is facilitated by these social ties and constant contact . . . trust-based governance seems easy to sustain when it is spatially clustered.
>
> (1994: 386)

Work in this second analytical vein leads Amin and Thrift (1994: 12–13) to suggest that attention in the literature on industrial agglomerations has increasingly turned from 'economic' reasons for the growth of new industrial agglomerations to 'social' and 'cultural' reasons. The latter include: levels of inter-firm collaboration; institutional support for local businesses; and structures encouraging innovation, skill formation and the circulation of knowledge.

Thus, while some regions may depend on dynamic urban processes, others congeal around reciprocal relations and the socio-economic networks that straddle both urban and rural areas. As we show in Chapter 2, economic regionalisation appears to be emerging in the British countryside at the present time. As agriculture moves to being only a minor part of the rural economy, so manufacturing and service industries come to provide the bulk of rural jobs. The economies of rural areas therefore become increasingly tied into regional patterns of development (Ward *et al.*, 2001). Rural areas located within the ambit of large cities tend to find themselves caught up in the movement of high-tech industry and financial services into the peri-urban fringe. These locations currently hold economic growth rates in excess of those in the inner city and the incorporated rural areas can be rendered extremely dynamic by their inclusion in 'new industrial spaces' (Gillespie, 1999). However, such tendencies are not universally evident for rural areas located a long way from the main urban markets often experience great problems in generating any kind of dynamic growth. Yet, even these areas have seen changes in the structures of their economies (Day *et al.*, 1989). While some of the new activity is attributable to the location of branch plants and call centres (so that the work provided is of the low-skilled kind), it can stem from firms and networks that utilise local resources in innovative ways (Moseley, 2000). Much of it is also driven by the changing geography of consumption, as residential dispersal, retirement to the countryside and rural tourism draw in supporting services.

Political regionalisation

Economic distinctions between regions can be further enhanced by political forms of regionalisation as political institutions act to support regional economies. In a helpful overview, Keating (1997) assesses political forms of regionalisation under three main headings: 'functional integration'; 'institutional re-structuring'; and 'political mobilisation'. The first of these – 'functional integration' – is most evident in the accounts of economic and technological change mentioned above, where new spatial forms are being delivered via rounds of economic re-structuring. Such re-structuring re-defines the nature of the capitalist space economy with the consequence that, in Keating's view,

> economic development and the insertion of territories into the global economy depend on specific characteristics of territories. So modern developmental policies put more emphasis on indigenous growth, or the attraction of investment by qualities linked to the region such as environment, the quality of life or trained labour force, rather than on investment incentives provided by the central state.
>
> (1997: 384)

In the wake of these economic changes, Keating argues, the state is forced to build up its own regional capacities.

'Institutional re-structuring' refers to the decentralisation of governmental institutions. This may come about through a desire to enhance state efficiency through devolving the 'less gratifying functions' of the central state to the regional tier (ibid.: 384)[6]. Decentralisation may also occur in response to regionalist or local political demands. The activities of political movements which seek to consolidate state power at the regional level Keating calls 'political mobilisation' (ibid.: 388). In a variety of different ways, and for a variety of different reasons, political movements seek greater regional autonomy within or from the nation state. According to Keating

> there are integrative regionalisms, seeking the full integration of their territories into the nation and the destruction of obstacles to their participation in national public life. There are autonomous regionalisms seeking a space for independent action; and there are disintegrative regionalisms, seeking autonomy or separation.
>
> (1997: 389)

There is, then, no one model of political mobilisation around the region. Summarising the various trends, Keating says:

> There is thus a re-composition of political space in Western Europe, in which regions are emerging in two senses. They are political arenas, in which various political, social and economic actors meet and where issues, notably to do with economic development, are debated. At the same time, they are constituting themselves as actors in national, and now European, politics, pursuing their own interests.
>
> (ibid.: 390)

In Chapter 2 we show how current changes in the governmental structure are promoting political forms of regionalisation in rural England. For much of the post-war period, rural areas in the UK have found themselves under the sway of a centralised British state, one that tended to govern the countryside via agricultural policy. Crises in the agricultural sector, together with an urban–rural shift in manufacturing and services, have acted to dislodge this approach so we now witness the emergence of more regionalised rural development policies. The governance of regional rural development is still in its infancy but is being reinforced by regionalisation in the political sphere, notably devolution to Scotland and Wales and decentralisation, including the establishment of RDAs, in England. Both these innovations encourage government to assess the role of the countryside in a regional context and help to integrate rural concerns into regional decision-making processes. We discuss the new approach in some detail in Chapter 7.

Societal regionalisation

The processes of economic and political mobilisation described above also generate societal effects so that the region as a *social* arena is made more meaningful. In other words, the more entrenched the region becomes in economic and political structures, the more it becomes entrenched in social institutions also. Paasi (1991) suggests that the process of social regionalisation unfolds in discrete stages. First, the region begins to take shape as a recogniseable socio-political entity. It can then act as a focal point for social practices that were previously directed at other spatial scales. Second, the region begins to acquire a 'symbolic structure' which provides 'a framework for experience through which we learn who or what we are in society' (ibid.: 245). Third, regional institutions are increasingly strengthened so that individuals and organisations are 'socialised into varying regional community memberships' (ibid.: 246). And, finally, some form of regional consciousness emerges so that regional social and cultural forms gain much greater clarity and are no longer subsumed within local or national systems. In incremental fashion, the region becomes increasingly 'constitutive' for social actions (MacLeod and Jones, 2001: 251) and thereby becomes a 'collective representation' (ibid.: 678), one that cannot be reduced to either alternative forms of organisation or alternative modes of representation. More and more social activities are therefore carried out under a regional banner. In this way regionalisation becomes a self-reinforcing process.

In Chapter 2 we propose that this form of regionalisation may be starting to emerge in the English countryside. For instance, the urban–rural shift of industry is bringing English rural areas firmly within processes of economic regionalisation so that regional economic institutions are brought into being. These are accompanied by political institutions at the regional level and these institutions embrace rural as well as urban areas. Rural regions therefore gain greater degrees of 'institutional thickness' (Amin and Thrift, 1994). Moreover, the ever more extensive nature of social relations in rural England means that rural areas are now drawn into broad social structures that span urban and rural locations. The most obvious illustration of this is the counterurbanisation process in which household members live in rural areas but are employed in urban workplaces – their lives thus span the urban–rural divide. The growing number of counterurbanisers in the countryside increasingly serves to bind rural localities into regional social formations. Thus, trajectories of development in the counterurbanised countryside are dependent on trajectories of development in the region as a whole.

Our general argument here is that rural areas in England are being *regionalised*. However, we also note that regional areas are being *ruralised*: there is a rural component to the regional economy, to regional politics and regional social formations. This book is an exploration of this 'regionalised rurality' and it attempts to show how rural areas are being incorporated into emerging regional contexts. For instance, in Chapter 4 we examine how rural land use in South East England is shaped by the demands of the counterurbanising middle class. This social group – which is strongly present in the region – reacts against the buoyant regional economy by seeking to protect the environments of rural areas from further economic development. The alliances and partnerships that are built up to achieve such protection help to stabilise rural social relations in line with environmental values. These alliances also affect the functioning of regional institutions so that preservationism and other environmental objectives become apparent in regional policies. In the process, rural land becomes 'fixed' as an environmental asset in a dynamic economic context. In more remote rural areas, as we show in Chapter 6, this counterurbanising social formation may be very weak and therefore alliances and partnerships will be oriented not to protectionism but to economic growth. Local and regional social formations can therefore coalesce around rural development priorities and these are likewise adopted at the regional governmental tier. Yet other regional areas, illustrated by the case study presented in Chapter 5, may exhibit tensions between development and environment so that regional institutions and alliances compete to implement the differing priorities for the area. In all these ways, social groups struggle to define the shape of newly regionalised ruralities.

Networks and conventions: a theoretical framework for the differentiated countryside

While this brief overview indicates that rural areas are currently being incorporated into revised regional structures, it also indicates that the 'region' has a variable geometry: it emerges from new economic agglomerations that link firms and other institutions within industrial ensembles of varying shapes and sizes; it is demarcated by the new administrative regions as powers and functions are decentralised within multi-level governance systems; and it is pursued by political and social actors as regions are symbolically carved out of the confluence of territory and social relation. In all these ways, the term 'region' is made more meaningful and substantial. However, each process covers a different territory; thus, no overarching and final definition of 'the' region can be given. As Allen and colleagues (1998: 2) say, 'there is no complete portrait of a region' because 'regions only exist in relation to particular criteria'. We should therefore see the 'region' as being composed of multiple processes and relationships.[7]

The arguments reviewed in the previous section suggest that current changes in economic, social and political structures are yielding new institutional and organisational arrangements. In considering how these new arrangements might be apprehended within a single theoretical framework, the literature proposes that the economic, political and social relationships now demarcating discrete regions might be thought of as *networks*. According to Castells (1996: 470), 'networks are appropriate instruments for a capitalist economy based on innovation, globalisation and decentralised concentration'. As we have seen in the previous section, these various processes are thought to be provoking strong patterns of regionalisation at the present time. Thus, regionalisation would appear to go hand in hand with the rise of a 'network society' (ibid.). And we should expect this 'society' to have a marked effect not just on the urban 'hubs' of the global economy but on rural areas also.[8]

In thinking about the interaction between network society and rural territory it is worth briefly reviewing theoretical understandings of networks. In general, the term 'network' refers to a specific type of relation linking a defined set of persons, objects or events (Knoke and Kuklinski, 1982). While various definitions of networks can be used, all tend to refer to a discrete sets of linkages between actors or nodes; that is, the world may be made up of an almost infinite number of network linkages, but only some (e.g. particular economic, political, social or technological relations) will be of analytical interest (Breslau, 2000). It is therefore necessary to spell out the theoretical focus of any network study in advance in order to identify the network forms that will be investigated.

For instance, studies of economic regions focus analytical attention upon the 'interdependencies' between firms and other institutions because these links are assumed to provide contexts for effective economic action at the regional scale (Cooke and Morgan, 1993; Staber, 2001; Storper, 1997). In the political realm there is also growing interest in network forms and the power relations that bind them together. Recent studies of the policy process have recognised that 'old-style' government – that is, top-down, hierarchical decision making – is much less effective in carrying through state programmes and policies than forms of policy-making that work through alliances and partnerships (Rhodes, 1997). Likewise, contemporary social change is frequently apprehended using the network repertoire. In its sociological guise, network theory rejects the view that social action derives from societal-wide structures or norms; instead it builds its explanations from analysis of the patterns of relations that surround any social action or event (Emirbayer and Goodwin, 1994).

Although there are differing uses of the term 'network' in the academic literature, there is no doubting the growing popularity of the term. Urry (2000) has recently suggested that heightened academic interest in networks follows from the dissolution of endogeneous social structures, encased within strong nation states, as a result of globalisation trends. The fragmentation of contemporary societies means that disciplines such as sociology, which have been established upon the basis of stable national social structures, must seek out new objects of analysis using new conceptual tools and analytical frameworks. Urry therefore directs our attention to the flows of objects, information and peoples that run through and around national territories. Writing with Scott Lash, he argues that,

> The movement, the flows of capital, money commodities, labour, information and images across time and space are only comprehensible if 'networks' are taken into account because it is through networks that these subjects and objects are able to gain mobility. Whatever form of institutional governance is dominant, whether markets, hierarchies, the state or corporations, the subjects and objects which are governed must be mobile through networks.
>
> (Lash and Urry 1994: 24)

Networks and flows emerge from differing points of origin (global, national, regional, local arenas) and carry their cargoes over varied distances. Sociology must therefore direct its attention to the 'heterogeneous, uneven and unpredictable mobilities' that run through and around given social spaces (Urry, 2000: 38). Although he provides no clear definition of the term 'network', Urry claims that network analysis is a particularly useful means of approaching this mobile 'post-national' terrain.

In this book we consider the different ways in which network relations 'coalesce' with given socio-spatial formations in the countryside. Our interest here is in the differing networks that surround land development processes and in the way these networks shape land development outcomes. However, in investigating such networks, and the varied means by which they orchestrate processes of change in rural areas, we introduce a note of caution into Urry's argument: we propose that while networks might appear to be increasingly mobile, that is, increasingly disconnected from given spaces, they also act to reflect and refine broader spatial relations. We believe the process of 'reflection' and 'refinement' ensures a continuing interrelationship between networks and territorially distributed resources. It can therefore help to account for regionalisation processes in the countryside.

By considering the interaction between spatially distributed resources and fluid and mobile networks we propose that the network perspective might act as a form of 'middle-level' analysis, one that sits *between* broad structural changes on the one hand and the localised actions of individuals or groups on the other. In adopting this approach, we follow Breslau (2000) who argues that network theory must always be accompanied by broader modes of sociological investigation. He says, 'despite the utility of the network concept as a way of describing the relational constitution of entities, reality in general and social reality in particular cannot be derived from network connections alone' (ibid.: 302). In this view, the incorporation of actors into networks is partially determined by the structure of the broader social field so that only certain types of relations are seen as appropriate at any moment in time. Thus, we can assert that, while networks have the potential to forge new alignments and identities (Murdoch, 1997a), their scope is limited by broad socio-cultural assumptions about the types of entities that can legitimately be aligned with one another (Breslau, 2000).

In the following chapters we investigate the relationship between changes in social structures and the emergence of new network formations. We identify processes of re-structuring in modes of governance, the spatial distribution of economic activities, and the spatial composition of class structures and show that changes in the structural contexts surrounding the countryside are accompanied by changes in the dominant networks. In sketching out the relationship between these contextual features and particular network forms we concentrate upon the social and cultural predispositions that favour some network linkages over others. In so doing, we focus upon the way differing social actors – operating within networks – formulate and follow particular economic, political and social goals.

In order to link together the goals of actors, the activities of networks and shifts in broader structures we draw here upon the sociological theory of conventions. This theory has been used to examine the regionalisation of economic activity (Storper, 1997; Storper and Salais, 1997), the differing cultural conceptions that surround political practice in comparative contexts (Lamont and Thevenot, 2000) and the contrasting repertoires of development and environment that shape local disputes (Thevenot *et al.*, 2000). In general terms, it concerns 'the way persons are evaluated as moral or political agents and the way things are caught up in such evaluations' (Thevenot, 2002: 53). In our view this focus on evaluatory practices provides a potentially helpful means of comparatively assessing the activities of differing networks and allows us to link structural change to local actions in the differentiated countryside.

Conventions theory proceeds from the assumption that any form of coordination in economic, political and social life (such as network building) requires agreement of some kind between participants (Storper and Salais, 1997: 16). Such agreement entails the building up of common perceptions of the structural context. Storper and Salais describe such perceptions as,

> a set of points of reference which goes beyond the actors as individuals but which they nonetheless build and understand in the course of their actions. These points of reference for evaluating a situation and coordinating with other actors are essentially established by *conventions* between persons. . . . Conventions resemble 'hypotheses' formulated by persons with respect to the

> relationship between their actions and the actions of those on whom they must depend to realise a goal. When interactions are reproduced again and again in similar situations, and when particular courses of action have proved successful, they become incorporated in routines and we then tend to forget their initially hypothetical character. Conventions thus become an intimate part of the history incorporated in behaviours.
>
> (1997: 16)

Storper and Salais emphasise that points of reference are not imposed upon actors by some all-encompassing social order; rather they emerge through the 'coordination of situations and the ongoing resolution of differences of interpretation into new or modified contexts of action' (ibid.: 17). Thus, structural context (the 'point of reference') and social action (the 'on-going resolution of differences of interpretation') become combined in the associations that are stabilised between actors in networks.

Conventions effectively compose an arena of structured action in which social entities (both individuals and groups) utilise modes of evaluation in order to assess their own actions and the actions of others. The assessments are made in relation to notions of the 'common good', that is, whether 'what each person does meets the expectations of the others on whom he or she depends' (ibid.: 16).

Summarising the findings from an empirical study of convention types, Thevenot *et al.* (2000) suggest that the following repertoires constitute the main means of evaluating the 'common good':

- *market performance*, which leads to evaluations based on the economic value of goods and services;
- *industrial efficiency*, which leads to evaluations based on long-term growth;
- *civic equality*, in which the collective welfare of citizens is the evaluatory standard;
- *domestic worth*, in which value is justified by local belonging;
- *inspiration*, which refers to evaluations based on creativity;
- *reknown*, which refers to opinion and general social standing; and, lastly,
- *environmental justification*, which considers the ecological significance of human actions for the whole human/non-human collective.[9]

Each of these conventions depends upon 'a form of evaluation that singles out what is relevant' (Thevenot, 2002: 54). And what is relevant is evaluated in relation to some notion of the 'common good', assessed in terms of market performance, improvements in industrial infrastructure, welfare and equality, the strengthening of local ties, an improved environment and so on.

Social actors use the differing evaluatory repertoires in order to build relationships with other actors. Differing convention hierarchies are therefore consolidated in differing (economic, political and social) network formations. And as convention hierarchies become consolidated across the social field, so they shape the structural context in which networking activity unfolds. There is, then, an interactive relationship between network and structural context and this interaction constitutes the 'environment of action' in which further rounds of networking activity will be conducted. Each 'environment' establishes priorities for action in line with the conventions hierarchy. Differing network spaces therefore specify differing forms of action depending on the modes of evaluation that have drawn the involved actors together.[10]

While the theory of convention-based networks may seem rather abstract, we propose that analysis of differing conventions can help to provide insight into the two broad narratives of 'pastoralism' and 'modernism' identified in the Introduction (the narratives are, to some extent, disaggregated into convention types). And by linking conventions theory to network analysis we can suggest that the socio-material shape of any given rural area follows from the imposition of a particular conventions hierarchy, with rural land uses embodying the agreements established in the dominant networks. We can further argue that network agreements underpin processes of spatial differentiation in the countryside as conventions become consolidated in regionalised rural spaces. Spaces are therefore marked off from one another by practices of consensus building *within* networks and by economic, political and social conflicts *between* networks. In other words, the countryside emerges from processes of cooperation and competition and these processes give rise to an undulating geography of networks and conventions.

In the following chapters we investigate processes of cooperation and conflict within the land development process. We argue that as new economic, political and social networks emerge from broad processes of re-structuring so they make new developmental demands that shape the uses to which rural land is put. On the one hand, networks of social actors seek to implement conventions associated with local worth, environmental value and preservationism; on the other hand, economic networks push conventions of economic efficiency, market demand and rural development. The contrasting landscapes of rural regions can therefore be seen as reflecting the priorities for action specified by the networks and the way these priorities are imposed in land development processes and outcomes.

Conclusion: the structure of the book

As we have explained, the overall aims of this book are to consider how economic, political and social changes are conspiring to 'regionalise the rural' and to examine how these trends place differing rural areas on contrasting development trajectories. In Chapter 2 we sketch in the structural context and look in some detail at the main political, economic and social trends that are shaping patterns of rural development. In particular, we consider national post-war policies and argue that these have now given way to a set of more regionalised approaches. In Chapter 3 we go on to examine the theoretical explanations of countryside change that have been proposed in recent years. We identify theories of political, economic and social development in rural areas and describe these in terms of a broad movement from a 'national' orientation to a focus upon regional and local differentiation. We show that this movement can also be viewed in terms of a shift in dominant structures as a result of new network configurations emerging in rural areas. We outline how these new networks have been perceived in discussions of rural politics, the rural economy and rural society.

Having outlined recent theoretical developments, we then turn in Chapter 4 to the first of our study areas – Buckinghamshire in the South East of England. This county – which was the subject of an earlier work, *Reconstituting rurality* – lies close to London and illustrates how rural areas can be caught up in dynamic urban contexts. It also shows how changes in the social structures of rural areas are closely linked to the networking activities of new rural residents. In this case, we argue that the assertion of rurality in the context of a developmental countryside is expressed most forcefully through the planning system as counterurbanising residents attempt to resist any further development. They therefore build strong associations in the planning arena in order to preserve existing rural land resources. In so doing, these actors produce an 'environment of action' that is strongly predisposed against further development while remaining conducive to further middle-class in-migration. Yet, the county's location ensures that development pressure is almost continuous. Thus, resistance to development is an ongoing process, one that is deeply entrenched in the social and political formations that surround planning.

In Chapter 5 we examine Devon, a county which, because it is located some distance from any major conurbation, still retains a rather traditional economic, political and social structure. However, Devon's location in the 'south' means that counterurbanisation rates are high as people move into the area either to set up new businesses or to retire. Thus, many of the same pressures that we witness in Buckinghamshire are becoming evident in the county. In particular we find that new residents tend to form anti-development alliances in order to counter the traditional political strength of the pro-development constituency. The result is a considerable amount of conflict in local politics as the two local groups confront one another.

In Chapter 6 we turn to Northumberland, a county in the North East of England. Here an even more traditional social structure remains in place, held together in part by the landed estates that have survived from the nineteenth century. The owners of these estates retain a key influence over patterns of rural development. And because counterurbanisation remains constrained, the environmental concerns usually expressed by the incoming middle class are much less central to political practice than in the previous two study areas. We can thus see the consolidation of a much more developmental 'environment of action' in Northumberland, one built around traditional concerns for rural socio-economic development set within a localistic context.

In Chapter 7 we draw out a comparative analysis from the three case study chapters in order to elaborate a number of general findings. We then set these 'findings' within the context of policies for the differentiated countryside. We show that contemporary policy changes are taking a regionalised form and these carry significant implications for how the differentiated countryside is governed.

In Chapter 8 we reflect upon the dynamics of differentiation and outline possible futures for the differentiated countryside.

In short, we seek to analyse here *differentiation* in the context of *regionalisation*. We aim to provide an account which sets out the main causes and consequences of these two interlinked processes and which allows us to consider how they work to shape the modern countryside. We undertake this task by examining the interaction between economic, political and social processes in the context of rural land development. And in analysing land development processes in differentiated rural locations we aim to assess the changing relationship between the narratives of pastoralism and modernism identified in the Introduction. We seek to show that these narratives can only be adequately assessed if they are situated in the diverse and complex developmental spaces that now comprise the differentiated countryside.

2 Regionalising the rural

Introduction

The notion of a differentiated countryside arises from a belief that the national rural space consolidated after the Second World War has now given way to a number of increasingly distinct rural *spaces*. In the early post-war period, dominant economic processes seemed to be orchestrating a homogenisation of the countryside. In contrast, the main processes currently operating appear to be producing divergent socio-economic formations in rural areas. These formations, we shall argue, are best seen as consolidating at the *regional* level. In later chapters, we examine a number of case studies from different regions and compare these cases in order to provide an overview of the differentiated countryside. In this chapter, we describe the shift from a national rural space to a differentiated set of regional formations, and we concentrate, in particular, on the changing role of public policy.

First, we show how the assumptions that underpinned policy in the wartime and post-war period relied upon a discourse which privileged the 'national' over the 'local' and the 'regional'. This was most evident in the sphere of agricultural policy, which dominated the rural policy agenda at this time, but can also be detected in conservation and environmental policy. The requirements of post-war economic development demanded an agriculture that played a central role in national re-construction. As we shall see, this role was driven by a conception of the 'national farm' in which all agricultural areas were expected to contribute to the greater national 'good'. The construction of the 'national farm' involved the elimination of any local and regional variations that did not contribute to the national productivist regime (see Lowe *et al.*, 1993).

During the last quarter of a century or so a variety of pressures have combined to provoke a re-assessment of the post-war settlement and many of its underlying assumptions. In particular, the central place of agriculture in rural policy has come to be widely questioned. There are two main reasons for this. First, agriculture is no longer the main economic activity in rural areas. With the inexorable decline in agricultural employment, and in the wake of an urban–rural shift in manufacturing and services, rural economies now look much more like urban economies, at least in terms of their sectoral composition. Second, agriculture has not been the steward of the rural environment, as the post-war settlement had assumed it would be. Rather, farming has wrought profound and what are widely seen as destructive changes on rural nature. These two factors, when allied to the ever-increasing cost of agricultural support and successive food and farming crises, have provoked a significant re-thinking of agricultural policy.

Having examined the post-war agricultural settlement and its associated planning policies, the chapter then goes on to consider two trends that have done much to force a re-assessment of policy-making in these two policy sectors: the urban–rural shift of manufacturing and services and the movement of population from urban to rural areas. These two trends – which are inter-linked – appear to fatally undermine the association of the *rural* with the *agricultural* for they imply that the rural economy is a great deal broader than the agricultural economy and that rural society has only a tenuous connection to the agricultural sector. We examine the forces that underlie the two trends and show how they are combining to disaggregate national rural policy into a set of regionalised approaches.

The emergence of rural policy

For most of its history the British countryside has been characterised by considerable diversity. Differences in climate, soil type and terrain, allied to variations in culture, cuisine, dialect, custom and tradition, ensured that until the mid-nineteenth century forms of life in rural areas were quite heterogeneous in nature (Murdoch, 1996). However, this diversity was undermined, first by improved transportation links between previously distant rural places, and second by the shift of industry, and thus population, into towns following the industrial revolution (Lowe and Buller, 1990). As a result, rural economies became specialised in agricultural production. Moreover, agriculture itself was driven by an increasingly commercial approach, which stimulated technological change and standardisation throughout the sector (Wormell, 1978).

Since the end of the 1930s, the commercialisation and industrialisation of agriculture have been promoted by the state. Before then, the state sustained only a fitful interest in the industry: although agriculture was seen to have strategic importance during wartime, governments were less willing to lend support during times of peace (they preferred instead to concentrate on ensuring the provision of cheap food). With the outbreak of the Second World War, however, a much more robust structure of agricultural support came into being (one that, in significant respects, lasts until this day). The outbreak of the war brought state control of farming as the whole sector had to be mobilised to maximise domestic food supply (Murray, 1955). This was imperative in order to diminish dependency on food imports, which were both vulnerable to enemy action and took up space in naval convoys needed for military and other supplies. As Kirk (1979: 47) puts it, agriculture 'was made into a sort of ward of the state, much like a nationalised industry'.

A useful starting point for any analysis of the post-war agricultural settlement is the 1942 Report of the Scott Committee on Land Utilisation in Rural Areas. The Report's importance stems from the way it 'brought together for the first time, in a formal and authoritative statement, the separate strands of land use policy and the requirements of agriculture' (Sheail, 1997: 390). Significantly for the arguments pursued in this chapter, the Committee adopted the view that 'the most certain way of protecting the wellbeing of rural communities and enhancing both the amenity and recreational value of the countryside was to retain the existing area of farmland in productive and profitable use' (ibid.). Thus, the Scott Committee recognised that agriculture should constitute the major national interest in rural areas.

The government of this national interest entailed state tutelage of agriculture, with two main components. First, as a Government statement of 1940 put it, the state should be committed to a 'healthy and well-balanced agriculture as an essential and permanent feature of national policy' (quoted in Sheail, 1997: 393). This commitment implied financial and technical support for farmers. Second, the town and country planning system, which the Scott Committee believed had failed agriculture in the interwar years by allowing too much land to be lost to urban sprawl, should protect rural land as a space in which agriculture could flourish.

With its emphasis on the need for a strong and unified *national* system of development control in rural areas, and its concern with investment and stability in the agricultural sector to meet *national* food requirements, the substance of the Scott Report fed directly into both the 1947 Agriculture Act and the 1947 Town and County Planning Act. These two acts – the 'twin pillars' of post-war rural and agricultural policy – were directly related to one another. The Agriculture Act sought to promote 'a stable and efficient agricultural industry capable of producing such a part of the nation's food and other national produce as it is desirable to produce in the UK'. The Planning Act's main function in rural areas was to protect agricultural land from urban development. As Hodge (2000: 94) puts it, at the time 'it was strongly believed urban development represented the major threat to the quality of the countryside whose character and quality depended upon its maintenance by a prosperous agricultural sector'. Thus planning was given a strongly protectionist function (see also Hall *et al.*, 1973; Cullingworth, 1997). While both pieces of legislation allowed a degree of local sensitivity, they situated diverse rural areas within a single, coherent national policy framework which sanctioned comprehensive state intervention in the rural domain. They effectively spelt the end of a centuries-old tradition in Britain which specified that the conduct of rural areas and rural affairs was properly a matter of local judgement and discretion.

Post-war agricultural policy

The 1947 legislation reflected the primacy that agriculture had enjoyed during the Second World War. Yet, to those charged with mobilising it, the farming world appeared hopelessly backward and fragmented and the aim of policy was to introduce some measure of standardisation in the sector. Looking back, one of the senior civil servants who directed the process commented:

> In translating the paper plans of Whitehall into concrete action on some 300,000 farms . . . the [Agriculture] Department was well aware that it would have to cope . . . with the wide diversities of farming practice which were the inevitable result of the differences in soil and climate in various parts of the country.
>
> (Winnifrith, 1962: 27)

The modernisation of agriculture therefore seemingly required an extensive rationalisation of the structure of farming in order to make farmers effective partners with the state.

Rationalisation was furthered by a whole variety of measures. First, the system of guaranteed prices introduced during wartime was continued. The effect of this system was to reward output; thus those producers whose production was greatest gained the most in terms of financial support (Lowe *et al.*, 1986). In turn, the nature of support provoked a consolidation of holdings so that farm size began to increase (Allanson, 1992). Second, the state sought to increase agricultural efficiency and competitiveness through the provision of business and improvement grants. Business grants were aimed at promoting entrepreneurial thinking among farmers and stimulated the use of accountancy and other business management techniques (Murdoch and Ward, 1997). Improvement grants encouraged farmers to undertake drainage and other land improvements to boost productivity (Wormell, 1978). Third, through such means as the criteria for grants and the targeting of advice, the policy promoted a particular model of the professional farmer – someone who was full-time, formally trained and with a progressive outlook, able to absorb technical advice and new practices (Brassley, 2000). Fourth, farmers were encouraged to adopt common business and husbandry practices by the state-funded National Agricultural Advisory Service (Murray, 1955). All these measures promoted the homogenisation of farming practice, with the avowed aim of 'efficiency' and 'increased productivity' (Bowers, 1985).

In this fashion, British agriculture became conceptualised as a national sector of production, with aggregate output of key commodities the guiding principle in its joint management by state and farmers (Murdoch and Ward, 1997). Nowhere was this more evident than in the Annual Price Review, a forum that had been instigated in 1943 and which became the 'font' of national agricultural policy after it was given a statutory basis by the 1947 Agriculture Act. The Review procedure fell into two parts. First, agricultural ministers were required to consider each year the general economic condition and prospects for the 'industry', consulting any bodies that appeared to them to represent producer interests, which in practice meant the National Farmers' Union for England and Wales (NFU), the National Farmers' Union for Scotland and the Ulster Farmers' Union.[1] Second, ministers would then decide what changes were required for the ensuing year in the level and distribution of guaranteed prices and the various production grants.

The Review required the collection and collation of an enormous amount of statistical material. The economists of the Agricultural Departments and the farmers' unions assembled much of the information prior to the Review deliberations, including: estimates of the aggregate net income of the industry, and its distribution by type and size of farms; changes in transport, labour and other costs; a comparison between farm incomes and other classes of incomes; the size of the subsidy bill in total; and the amount to be spent on different commodities and production grants. All this data was used to compile a picture of the 'national farm'. Once this picture had been constructed then policies could be put in place which encouraged and coerced farmers to play their part in boosting production (see Short *et al.*, 2000).

The 'national farm' was underpinned by an extensive institutional and administrative apparatus. The key agencies were the Ministry of Agriculture and the NFU who worked together to extend and apply the national policy, persuading a multitude of independent farmers to adopt modern farming practices in return for state support. The relationship between the Ministry and the Union within the context of the Annual Review was pivotal to the formulation and implementation of this policy (as has been well-documented for example by Allen, 1959; Bowers, 1985; Cox *et al.*, 1986a; 1986b; 1987; Grant, 1983; Self and Storing, 1962; Smith, 1989; Winter, 1996). Once it had been 'brought inside' the policy process the NFU could claim to be the exclusive 'voice' of the farming industry, further boosting its membership which reached 200,000 by 1949 (Murdoch, 1988). It thus played a key role in 'constructing' farmers' interests in such a way that they fitted the requirements of the Annual Review. Likewise, the post-war growth in state intervention in agriculture was paralleled by a growth in the administrative structure of the Ministry. By the early 1950s the Ministry of Agriculture had established an extensive local structure and had grown to be six times the size it had been in the late 1930s.

The subsidy and grant system in operation during this period played a major part in the achievement of these goals. By 1969 agricultural output stood at nearly twice its pre-war level while the number of farmworkers employed on farms had more than halved (Lowe *et al.*, 1986). The UK's entry to the European Community and its Common Agricultural Policy (CAP) in the 1970s did little to change the thrust of state intervention. The Treaty of Rome, which had laid the foundation for the European Community in 1962, stipulated that agricultural policy should 'increase agricultural productivity by promoting technical progress and by ensuring the rational development of agricultural production and the optimum utilisation of the factors of production, in particular labour' (quoted in CEC, 1987), an aim that neatly dovetailed with Britain's own agricultural policy aspirations (Fennell, 1979).

Entry to the European Common Market therefore did little to alter the underlying aims of agricultural policy, a point that became especially apparent when the UK government published *Food from our own resources* in 1975. This document, which set a course for policy over the next five years, stated that, for balance of payments reasons, 'the Government takes the view that a continuing expansion of food production in Britain will be in the national interest' (MAFF, 1975: 4), a theme that was reiterated in *Farming and the nation* published four years later (MAFF, 1979). The policy of 'productivism' and the achievement of national self-sufficiency in food that could be grown in Britain remained in place.

However, by the early 1980s problems with the operation of the CAP were becoming manifest. It was not just UK agriculture that was expanding production; member states throughout the European Community were doing just the same. The Community was faced with surpluses in a number of the main agricultural commodity sectors (such as sugar, wine, cereals and milk). Consequently, expenditure under the CAP – to maintain domestic prices and to store, and subsidise the export of, surplus production – rose rapidly (at a rate of 25 per cent each year in the late 1970s and early 1980s). In the mid-1980s a number of measures were introduced in an attempt to curb overproduction, the most significant of which were milk quotas and set-aside for cereals. These measures attempted to constrain production levels while also maintaining farm incomes. In this respect, they continued the traditions of the post-war settlement and protected farmers from market forces (Fennell, 1997; Gardner, 1996). However, they also began to differentiate farmers from one another (for instance, between those with quotas and those without) and such differentiation had a geographical effect as the quota that was transferred between farms tended to drift from west (e.g. the small farms of Wales) to east (e.g. the larger dairy farms in England) (Murdoch, 1988).

Concern about the cost of the CAP was bolstered by a recognition, again from the early 1980s onwards, that the policy had not only fundamentally altered agriculture but the countryside too (see, for example, Bowers and Cheshire, 1983; Lowe *et al.*, 1986; Shoard, 1979). Hodge summarises the environmental changes wrought by the post-war settlement as follows:

> In the 40 years following the Second World War, about 95 per cent of lowland meadow was lost, 80 per cent of chalk downland, 60 per cent of lowland bogs, 50 per cent of lowland marsh and 40 per cent of lowland heath . . . The length of hedgerows declined from 495,000 miles in 1947 to 386,000 in 1985.
>
> (2000: 100)

With growing recognition and concern over the damaging consequences of productivist agriculture, it became clear that agriculture could not be relied upon as the guardian of the rural environment.

This was a matter of considerable contestation between farming and conservation interests as the latter sought a more direct influence over agricultural production and farm management decisions (Potter, 1998). Following a series of controversies, in which agriculture was seen to be damaging scarce natural resources, areas that were suffering acute environmental problems were highlighted and specific management agreements were negotiated. A range of area-based policies emerged – Sites of Special Scientific Interest, Environmentally Sensitive Areas (ESAs), Nitrate Sensitive Areas, Countryside Stewardship Schemes and so on – and these gradually re-introduced regional and local diversity into the national policy framework (Lobley and Potter, 1998).[2]

Recognition of agriculture's environmental problems, allied to spiralling budgetary costs and opposition from the EU's trading partners to the dumping of surplus agricultural production on world markets, led to sustained efforts to reform the CAP, although achieving effective reform has proved a protracted and piecemeal process (Kay, 1998). In 1985 the European Commission published *Perspectives for the common agricultural policy* which set out the general direction that reform should take. It warned that:

> Unless the Community succeeds in giving to market prices a greater role in guiding supply and demand within the agricultural policy, it will be drawn more and more into a labyrinth of administrative measures for the quantitative regulation of production.
>
> (CEC, 1985: 4)

In an attempt to steer clear of 'labyrinthine' solutions to farming's problems (such as quotas), the Commission subsequently put forward a number of market-oriented packages and solutions. The so-called 'McSharry' reforms of 1992 cut support for some products and introduced compensation payments to farmers. They also established a set of social and environmental 'accompanying measures'. However, while these reforms promised a re-shaping of the CAP, change was incremental rather than revolutionary (Potter, 1998; Brouwer and Lowe, 2000).

Dissatisfaction remained over the budgetary cost and social and environmental side-effects, but there were also pressures from external sources, notably the proposed expansion of the EU into Eastern Europe and the demand by the World Trade Organisation (WTO) (formerly GATT) that agricultural markets in the EU be further liberalised (Barclay, 1999). *Agenda 2000*, published in 1997, comprised the Commission's response to these pressures (CEC, 1997). It suggested lower prices for a variety of products, greater decoupling of support from production, and more emphasis on rural environmental and rural development measures. The proposals signalled a shift in resources away from price support and towards a more integrated rural policy approach (Lowe and Ward, 1998).

Negotiations around the Commission's proposals culminated in a deal at the Berlin Summit of March 1999. This settlement continued the process of reform, but at a much slower pace than had originally been envisaged by the Commission. The various compromises agreed on price reductions did little to close the gap with world prices and committed member states to continuing high levels of agricultural spending. The main innovation was the adoption of the Rural Development Regulation which emerged from the simplification of nine former measures associated with agricultural structures and environmental schemes. The Regulation was made up of the 'accompanying measures', which included agri-environment support (such as Environmentally Sensitive Areas, countryside stewardship and organic aid) and Less Favoured Area support, and 'non-accompanying measures' which included investment in agricultural holdings, training, marketing and processing grants, and general rural development support. The Rural Development Regulation has been hailed by the Commission as the 'second pillar' of the CAP (the first pillar being the commodity supports) and it is claimed that, with its introduction, the institutional framework for re-orienting the direction of European policy from an agricultural-sectoral orientation to a rural-territorial one is finally being put in place.[3]

These reforms are giving rise to a more regionally diverse approach to programming the CAP. Not only will the general shift to a more market-sensitive policy lead to greater distinctions between farmers (i.e. between those that can 'compete' and those reliant on continuing state support), but the environmental schemes that now comprise a growing part of state expenditure will also distinguish those with important environmental assets from those without. The *Agenda 2000* reforms allowed for a territorially differentiated approach to the programming of the Rural Development Regulation. In the UK it has been introduced on a regionalised basis so that Scotland and Wales now administer their own Rural Development Plans (under the devolved governments of those countries) while the English plan is divided into nine administrative regions. It would seem that as rural development becomes a more important aspect of agricultural policy, so that policy is likely to become more regional in nature. We return to this theme in Chapter 7.

Post-war planning policy

Agricultural policy in the post-war era has been accompanied by a national planning policy, one that has been charged with protecting the countryside from urban development. In fact, agricultural policy and planning policy have complemented one another to a remarkable degree over the last fifty years or so. Thus, the post-war planning system must be seen as a central component of the 'productivist' regime. In this section we assess the role that planning policy has played in binding rural areas into a national mode of regulation.

Land use planning in the post-war period combined two main schools of thought that had emerged within the planning profession during the interwar years. The first argued that planning should be oriented to the preservation of the countryside, to protect agriculture, farmland and the rural landscape. This argument was advanced by interest groups such as the Council for the Preservation (later Protection) of Rural England (CPRE) and the Garden Cities and Town Planning (now Town and Country Planning) Association, and by leading planners, such as Patrick Abercrombie (a founder-member of the CPRE). The preservationists proposed that planning should act to prevent suburbanisation, urban sprawl and the use of rural land for urban purposes (Matless, 1998), a mode of thinking that was to the fore in the Report of the Scott Committee (which included Abercrombie and another leading CPRE activist, Dudley Stamp). Second, following the Barlow Report of 1940, it was suggested that planning could help to re-direct economic resources in order to overcome high levels of unemployment.[4] Thus, planning should interact with regional policy to re-distribute economic activity in a more geographically balanced fashion (say, by restricting growth in buoyant locations like the South East of England and promoting development in depressed industrial areas such as South Wales and the north of England) (Wannop and Cherry, 1994).

As Reade shows, rural land protection came to be accepted as a central goal of planning because it was strongly linked to so many other aspects of planning policy:

> To solve the housing problem, made acute by six years of war, new towns would be created, separated from the big cities by wide green belts, the latter having obvious appeal to rural preservationists. Eventually, the growth of these new towns would permit reduction of excessively high residential densities in the poorer quarters of the big cities, which could then be re-built to far higher standards, with more open space. Curbing the growth of the cities by means of the green belts would oblige any industries within them which sought to expand to re-locate either in new towns or in the depressed regions. . . . As for the countryside, its character would be zealously protected, and wherever possible, towns would have clean-cut sharp edges. Rural life would be re-vitalised, but any industry permitted in the countryside itself would be small scale, linked with agriculture as far as possible, and appropriate to the rural scene.
>
> (1987: 50)

This all-embracing, holistic policy programme was clearly ambitious. It also inferred that a strong regional dimension would exist within the new national planning framework. However, the Conservative governments of the 1950s came into office keen to dismantle war-time controls over industry. They were particularly averse to the regional steering of development and therefore downgraded the regional aspects of planning policy. Yet, they remained attuned to the need for urban re-construction and the concomitant concern to prevent urban sprawl. Thus, as economic and regional planning withered, the emphasis increasingly shifted to the implementation of a physical planning approach within a national system of regulation.

While the 1947 Town and Country Planning Act seemed to align the need for rural protection and the re-direction of economic activity to underdeveloped regions, in practice the twin goals of urban containment and the protection of agricultural land came to define the operation of the national planning system. Hall *et al.* (1973) summarise the objectives of post-war planning as:

- urban containment,
- protection of the countryside, and
- the creation of self-contained and balanced communities.

These three goals combined to ensure that a divide between the urban and the rural was established within planning policy and that planning in rural areas was mainly conducted *around* agricultural policy. With agriculture itself exempt from planning control, rural planning meant, in effect, planning the country towns and protecting agricultural land from development.[5]

One indication of the continuing importance of the urban–rural divide was the status of the green belt policies that had been gradually introduced in the early post-war period, notably in the city region plans of the 1940s. This approach was reinforced in an official circular of 1955, which provided for green belts to be incorporated into development plans. Many country areas around provincial cities responded and between 1955 and 1960 no fewer than sixty-nine sketch plans of proposed green belts were submitted to government. By 1963, 5,585 square miles of England were designated as green belts (S. Ward, 1994: 163–4). Once established, green belts became key tools of rural protectionism (Munton, 1983). This is what had been intended. As Henry Brooke, at the Ministry of Housing and Local Government, said in 1960:

> the very essence of a green belt is that it is a stopper. It may not all be very beautiful and it may not all be very green, but without it the town would never stop, and that is the case for preserving the circles of land around the town.
>
> (quoted in S. Ward, 1994: 164)

Yet, the extension of green belts did not prevent urban dispersal. At this time, many new private, housing estates were being established that appealed to the expanding ranks of the car-owning middle class (Clapson, 2000). These estates were mainly situated beside smaller towns in outer metropolitan areas within or beyond green belts, although some took the form of peripheral developments around small and medium-sized cities without green belts. This locational geography reflected, in part at least, government policy. For instance, the 1952 Town Development Act enabled small country towns to be expanded as rural local authorities entered into agreements with large metropolitan authorities in order to take 'overspill' population. The Conservative Government agreed to fund many of these schemes in order to channel population into certain demarcated growth centres. Rural planning therefore focused on the managed expansion of these and other growth centres and the protection of other rural areas and settlements (Cloke, 1983).

Nevertheless, concerns about the impact of urban sprawl on rural areas continued throughout the 1970s and came to a head most famously with proposals for private-sector new towns or 'new settlements' in the early 1980s. Encouraged by an apparent official sanctioning for the building of new villages and towns in the countryside, a group of house builders came together within one company – Consortium Developments Ltd – with the express intention of creating such settlements. During the mid-1980s the company put forward a number of schemes for private new towns, each designed to accommodate at least

5,000 inhabitants, located in the South East of England. Initial developments of this type, such as Lower Earley near Reading, seemed to indicate that they were a private-sector means of dispersing population away from the cities.[6]

In order to facilitate the building of new settlements in the countryside, Mrs Thatcher's Government sought ways to amend rural planning restrictions. However, such moves unleashed a storm of protest from environmental groups and residents of the rural South East, many of whom would normally be Conservative Party supporters. Feelings came to a head when in 1989 the then Secretary of State for the Environment, Nicholas Ridley, a free-market ideologist, professed himself minded to allow one new settlement – Foxley Wood in Hampshire – to be built. In the European elections of that year, the Green Party polled its highest ever share of the vote (15 per cent), pushing the Conservative Party into second and third place across much of southern England. The Party did particularly well in the South East, for instance taking 19 per cent of the vote in the Thames Valley constituency. In the words of one commentator,

> many Conservative voters living in outer-city and urban-fringe areas wanted a planning system that would stop their areas being changed by large-scale developments. However much they might approve of the enterprise culture in other walks of life, they did not want it in the form of opportunistic developers shaping their immediate home environments.
>
> (S. Ward, 1994: 207)

Following the 1989 European elections, rural protectionism was given a new emphasis in planning policy and the pro-development Nicholas Ridley was replaced as Secretary of State for the Environment by Chris Patten, a much more pragmatic politician. The thrust of policy now very quickly changed and the power of local planning authorities was re-asserted in the 1991 Planning and Compensation Act which stipulated that planning decisions should be made in accordance with the plan unless other material considerations indicate otherwise.[7] Moreover, plans were now to apply to the whole local authority area. Thus, for the first time rural areas were brought fully under the development plan, a measure that arguably strengthened rural protectionism in planning (see Marsden *et al.*, 1993; Cullingworth, 1997).

Reflecting on the evolution of the planning system over the period between the mid-1940s and the mid-1990s, Stephen Ward (1994: 267) remarks that 'restraint policies, particularly green belts, have been pursued with a sustained rigour that, though not absolute, has not been equalled in any other major aspect of planning'. Thus, in Ward's view, rural protectionism has been enshrined in planning policy so that the planning system has succeeded in erecting a divide between urban and rural areas across the national space. On one side of this divide – the urban side – development has unfolded in line with varied socio-economic demands; on the other side – the rural side – agriculture has dominated the rural landscape.

The renewed emphasis on rural protection continued into Tony Blair's first Labour Government (Allmendinger and Tewdwr-Jones, 2000). While this government was more likely than its Conservative predecessors to be sympathetic (in an ideological sense) to planning, it might also have been expected to stand up more robustly to the preservationist forces that have long ensured that planning protects the countryside (for instance, Labour has traditionally relied much less than the Conservatives upon shire county votes). However, in responding to concerns about the implications of increased demands for new housing, John Prescott, the Environment Secretary during Labour's first term, followed his Conservative predecessor in arguing that urban dispersal should be resisted and that most new houses should be built in existing urban areas. He actually increased the target for new housing on brownfield – as opposed to greenfield – land to 60 per cent and, to facilitate this, established an Urban Task Force (chaired by the architect Richard Rogers) to provide practical examples of how urban areas could be made to accommodate more houses (Urban Task Force, 1999). Spatial and environmental considerations were thus given even greater emphasis in Labour's planning policy (for an overview, see Murdoch and Abram, 2002).

Differentiation through a changing rural economy

For most of the post-war period, rural areas have been encompassed within a policy framework comprising nationally formulated conventions associated with the protection of agricultural land (for aesthetic and social purposes) and increases in agricultural productivity (for economic purposes). These national priorities have been reproduced in local rural areas so that an increasing homogenisation of rural space in both agricultural and planning terms has taken place. However, the scope and nature of national policy has recently begun to shift. This shift has been conditioned not only by changes internal to each of the main policy sectors – of agriculture and planning – but also by a set of external pressures that have emerged and built up in the post-war period (Marsden, 1992). Two such external pressures are of particular interest: first, the *urban–rural shift of manufacturing and services* and, secondly, *counterurbanisation*. These two interlinked trends have profoundly re-shaped the uses to which rural space is put. They have brought new demands upon the countryside and have changed the context in which these demands are assessed. In necessitating a broadening of the policy framework they have also required recognition of local and regional diversity within it. We look firstly at the urban–rural shift before turning, in the next section, to counterurbanisation.

The post-war period has seen an inexorable decline in primary sector employment with fewer and fewer people employed in agriculture, mining, quarrying and forestry. However, this decline was offset in the early years by increases in rural employment in the public services mainly as a result of growth in the national education and health sectors. Later, during the 1970s, it became evident that manufacturing and service industries were re-locating away from the conurbations and were providing new sources of employment in rural locations (Fothergill and Gudgin, 1982). By the 1990s, as a result of this urban–rural shift in employment, the employment profile of rural England, in aggregate terms at least, looked very much like that for the UK economy as a whole, with around 10 per cent employed in the primary sector, 20 per cent employed in manufacturing and 70 per cent in services (Butt, 1999; North, 1998; PIU, 1999). Such patterns of economic re-structuring have led to a profound reappraisal of rural economic change and the most appropriate policies to guide it.

The urban–rural shift was spearheaded by manufacturing industry. Between 1960 and 1987 the number of manufacturing jobs in England fell by 37.5 per cent but the number in rural locations rose by 19.7 per cent (Tarling *et al*., 1993: 34). In part, this trend was facilitated by changes in the structure of manufacturing industry in an era of 'post-Fordism' so that branch plants could be situated at some distance from company headquarters. Thus, rural areas could benefit from the inward movement of new plants attracted by low rents and wages (Urry, 1984). However, the urban–rural shift was also fuelled by the indigenous growth of small firms in rural areas, with new enterprise creation rates higher in recent decades in rural rather than urban locations (North, 1998; North and Smallbone, 2000). According to Keeble and Tyler (1995: 994) this reflects a new pattern of 'enterprising behaviour' in which 'rural settlements have been able to attract a relatively high proportion of actual and potential entrepreneurs, largely because of their desirable residential environmental characteristics'. Attractive and accessible rural areas thus offer 'competitive advantages resulting from the direct benefits of high-amenity living and working environment, greater labour force stability, quality and motivation, good management labour-relations, and lower premises, rates and labour costs' (Keeble and Tyler, 1995: 994–5)

A key component of rural economic success is therefore the preference for rural locations on the part of some of the most dynamic firms. Keeble *et al*. (1992) show that hi-tech businesses have been attracted to accessible rural areas in East Anglia (notably around Cambridge) and the Thames Valley (especially along the M4 corridor). Likewise business and professional services have favoured rural areas in their location decisions, with some of the highest rates of growth evident in accessible rural areas, especially in the south of England (Keeble and Nachum, 2002). Thus, while private service employment grew in the conurbations by 19 per cent between 1981 and 1996, it grew by 49 per cent in the towns and rural areas (Gillespie, 1999: 14; see also Turok and Edge, 1999). A feature of the growth of rural firms and employment is the positive effect of residential migration into rural areas, including the inward movement of new business owners and the induced growth of local services in response to residential expansion.

Yet, while a rather general shift in manufacturing and services has been taking place, it is clear that there continue to be very real differences between rural economies. On the one hand, many accessible rural areas are among the most economically advantaged in the country. They have become favoured locations for leading-edge industries (notably hi-tech and private service firms) and display higher than average income and GDP levels. On the other hand, remote rural areas and the former coalfields tend to demonstrate some traditional weaknesses – low wages, low skill levels, vulnerability and so on – and these are reflected in continuing low levels of income and GDP per head (Monk and Hodge, 1995).

Despite the undoubted buoyancy of rural areas as a whole, it is therefore clear that many rural places suffer a number of disadvantages including 'low wages, a limited range of job opportunities, a weaker skills base and poor access to education and training facilities' (Butt, 1999: 41–3). As a consequence, concern has been expressed that new industrial trends have served merely to reproduce the traditional economic weaknesses of some rural areas. So while the economies of peripheral rural areas have undoubtedly moved away from a reliance on primary industry, it is noticeable that places such as mid-Wales, Cornwall, Devon and Northumberland still have GDP levels way below the UK average (see, for instance, Countryside Agency, 2000).

In short, the urban–rural shift of manufacturing and services has resulted in considerable variation between rural areas. This is most marked in relation to the UK's broader economic geography, particularly the so-called 'north–south divide'. It has been a notable feature of economic change in the UK that the southern regions (the South East, the South West and East Anglia) have moved towards 'post-industrialism' (i.e. an economy dominated by services) much more quickly and smoothly than the north. Conversely, the 'deindustrialisation' that has taken place over the last two decades has affected the northern regions (Yorkshire and Humberside, the North West, the North and Scotland) much more disruptively than the south (Hudson and Williams, 1995). This has led to a decline in employment opportunities in the north while those in the south have expanded. Reviewing trends since the mid-1980s, Gillespie (1999: 11) points out that the northern regions are all continuing to lose employment, while the southern regions are still gaining. In accounting for this, he says:

> The principal differentiator of regional job performance is . . . a region's performance with respect to private services, which has become the motor of the British economy. . . . The dynamism in the national economy . . . is concentrated in the 'Greater South East' (now conventionally defined as the South East standard region plus the counties immediately adjacent). . . . A substantial disengagement from production activities has taken place, with vigorous growth in private services taking place in the South East but not in London. For the adjoining regions of the South West and East Anglia, a generally positive employment performance is apparent, as they have benefitted from spill-overs from the South East region. . . . The dynamism has clearly dissipated, or not reached, the more northerly regions, which all experience below average performance in generating private service jobs. Given the substantial loss of employment occurring in these regions, as elsewhere, in production activities, they are becoming increasingly dependent on public sector services to provide employment for their resident workforces.
>
> (ibid: 12–13)

Regionalisation thus follows from the integration of rural economies into their regional economies. Rural areas in the south of England inevitably benefit from the economic buoyancy of the whole area (based on service sector performance) and they become closely integrated into the regional economy, while rural economies in more peripheral areas tend to display the economic weaknesses of the regional context (e.g. an over-reliance on a declining manufacturing sector) (see Tarling *et al.*, 1993).

Differentiation through a changing rural society

The other main shift that has served to undermine post-war policy conventions is an increase in the rural population as a result of urban to rural migration patterns. For much of the twentieth century population has moved away from rural areas in search of employment in the cities. The continual shedding of labour from agriculture reinforced a process of urbanisation that had been set in train during the Industrial Revolution. However, the economic changes described in the previous section corresponded with a propensity on the part of more and more households to leave the city in search of a better life in the countryside. This search is termed 'counterurbanisation' (see Champion, 1989). Although counterurbanisation is an international trend, Champion (1994: 1504) argues that Britain has been in its 'vanguard': increasing affluence, along with changes in transportation systems, has allowed many people to combine country living with urban employment and more people are now moving away from cities than are moving into them (Boyle and Halfacree, 1998). The countryside is widely viewed as offering a better way of life than that available in the cities.

The first evidence of counterurbanisation came to light in the 1961 census and it showed that during the preceding ten years the population of metropolitan areas had grown by around 5 per cent but that this was matched by growth in non-metropolitan small towns and rural areas. Between 1961 and 1971 metropolitan growth was down to 3.5 per cent while the free-standing areas beyond the main metropolitan labour markets had increased their growth rates to 8.6 per cent (the rural component of the free-standing category had increased from −0.5 per cent in the 1951–61 period to +5.7 per cent in the 1961–71 period). Between 1971 and 1981 this trend continued, with metropolitan areas losing 2.3 per cent of their population while non-metropolitan areas increased their share by 6.0 per cent (again rural areas played a notable role, increasing their population by 9.4 per cent) (see Champion, 1994, for a summary of these trends). The 1971 and 1981 censuses thus showed clear evidence of counterurbanisation tendencies, with the fastest growing places being small cities and towns and remote rural areas (Champion and Townsend, 1990).

Between 1981 and 1991 the population of metropolitan Britain began to grow once more, by 0.4 per cent, yet the free-standing small towns and rural regions increased their total residents by 6.0 per cent. Again, it is notable that rural areas in both the metropolitan and free-standing categories increased their respective shares of population – by 7.3 per cent and 7.9 per cent. Champion (1994) concludes that population growth and net in-migration were relatively widespread across rural Britain over this decade although they were subject to regional variation with remote rural areas in East Anglia (+11.0 per cent) outstripping such areas in Scotland (+4 per cent).

Champion (1996: 14) likens urban to rural population movement to a 'cascade' in which 'each level of the hierarchy receives net in-migration from all higher levels and dispatches net out-migration to all lower levels in the manner of a full-scale "cascade", or indeed "general downpour"'. As population arrives into the 'inner city' (either through international in-migration or through rural to urban movements) so population is displaced to the 'outer city' and then down the settlement hierarchy. The central point to emerge from the 1991 census was that the cascade had a marked impact on some of the remotest rural areas in counties such as Devon, Dorset, Cornwall, Lincolnshire and Somerset. All metropolitan areas – with the exception of London, which managed to retain its population share – lost population during the same period. And evidence collected more recently seems to indicate that this trend has continued since the 1991 census. Champion *et al.* (1998) estimate that for the year 1990/91 there was a net movement of 80,000 people from urban to rural areas and they expect this rate to have been maintained throughout the decade. It has thus been calculated that the number of households in England's rural districts will increase by more than one million (that is, 19 per cent) between 1996 and 2016 (Countryside Agency, 2000; King, 2000).

In accounting for urban to rural population movement, three interrelated causal processes seem to be at work:

1 there are the particular choices that individuals make about residential location, choices which tend to express the rural ideal and some form of anti-urbanism;
2 changes in the structure of the economy – the ruralisation of economic activity – lead to a decentralisation of jobs and this encourages people to move to the countryside;
3 technological changes in transportation and communications, notably the increased use of the private car, the telephone and the internet, allow these locational choices to be expressed

(Lowe *et al.*, 1995).

In assessing the relative importance of the three causes we can speculate that it is perhaps the first that is the most powerful. For instance, of those questioned in the 1999 British Social Attitudes Survey, nearly 50 per cent said they would like to live in the countryside (of the sample, only 18 per cent said they already lived in a rural location). In justifying this response, 90 per cent of those questioned said the countryside provided a 'healthier environment to live in' (see Countryside Agency, 2000: 15). This bears out Fielding's argument that urban to rural migration is being driven by a form of anti-urbanism that regards cities 'as the sites of stress and conflict, and the countryside as the realm of harmony and sociability' (1990: 230). Increases in private affluence, associated with private car ownership and improved rail and road links, allow more and more people to express these motivations and choose a rural residence (Cross, 1990).

Given these underlying causes, patterns of counterurbanisation will be regionally and sub-regionally differentiated according to levels of private affluence, the proximity of rural areas to employment centres, the environmental and social 'qualities' of the area, the quality of transport links and the availability and cost of housing (Cloke, 1985). People who wish to commute will look for a house in an accessible rural location as rural living must be compatible with participation in the urban labour market. The ability to quickly leave the countryside for urban areas – and vice versa – is paramount, leading to complex urban/rural commuting patterns. If commuting is not a factor – either because the in-migrants are retired or are willing to seek work in their area of residential choice – then the rural characteristics of the location, linked to affordability concerns, will take precedence (Murdoch, 1998; Rogers, 1993).[8]

In assessing the impacts of counterurbanisation, Lewis (2000: 161) argues that the most important consideration 'is not necessarily population growth and net migration but rather the character of the flows into and out of rural settlements'. In other words, we need to look at who is leaving rural areas and who is coming into them. Perhaps the most striking feature of counterurbanisation is its class character. In general, studies show that those moving up the settlement hierarchy towards urban areas tend to be in the lower social classes, while those moving down the hierarchy towards rural areas tend to be young families and retired households from professional backgrounds (Boyle and Halfacree, 1998). This form of population turnover leads to a significant re-composition of rural society.

In short, rural society is becoming more 'middle class', and this social characteristic determines its changing preoccupations (Fielding, 1998; Murdoch and Marsden, 1994; Murdoch, 1995a). For instance, in rural politics rather less emphasis is placed on agriculture and rather more upon the rural environment and nature (Abram *et al.*, 1996; M. Winter, 1996). Lifestyles and cultural changes increasingly reflect the use of rural areas as consumption spaces (Cloke and Little, 1997; Marsden *et al.*, 1993; Ilbery, 1998). And within rural communities traditional rural activities are replaced by those which have a more 'post-modern' character (Bell, 1994; Murdoch and Day, 1998; Newby, 1985).

Moreover, the class character of rural space begins to 'fold back' on itself, mainly through the political activities of the new residents. Having moved out of urban areas in search of a better quality of life, counterurbanisers strive to protect that quality of life against external threats. In particular, middle-class residents place considerable value on the 'green' environments that surround their rural properties so that when proposals for new houses, roads, factories or waste tips come forward they often express their opposition vigorously (Murdoch and Marsden, 1994; Short *et al.*, 1986). And the more effective they become in opposing development the more the locale will reflect their aspirations. In other words, the rural area will take on a middle-class character (Phillips, 1998; Thrift, 1989).

However, we must be clear that the middle-class countryside is unevenly developed within the UK (Cloke and Thrift, 1990; Hoggart, 1997). As Lewis (1998: 138) points out, even during sustained periods of counterurbanisation there are rural districts that continue to lose population and to suffer problems of long-term decline. Such problems include large numbers of old people (the young tend to move out), declining community institutions and few employment opportunities. Moreover, the households moving into more remote areas will tend to differ from those moving into the more accessible countrysides. In the former it will be mostly retired people, as well as those seeking an alternative lifestyle away from the 'rat race', while in the latter it will be mostly commuter households (Green, 1997).

Counterurbanisation interacts with the urban–rural shift in employment to reinforce a 'regionalisation' of rurality. The economic attractiveness of accessible rural areas in southern and eastern England and the Midlands is linked to their social attractiveness, and here counterurbanisation is at its most advanced. In contrast, the declining former coalfield areas of the north and the Midlands attract few migrants. Likewise, the remote parts of the South West, East Anglia, the Welsh Borders and the north of England have low out-commuting rates. In turn, political pressures around development will be most pronounced in those areas that have witnessed the greatest amounts of counterurbanisation, with strong local coalitions against 'suburbanisation' formed on the back of earlier rounds of counterurbanisation (Savage *et al.*, 1992; Short *et al.*, 1986). In more remote rural areas, the traditional rural population will still tend to welcome greater job opportunities. People in such places are therefore more likely to be sympathetic to calls for economic development than calls for environmental protection. We return to these considerations in our case study chapters.

Towards a differentiated policy

The various economic and social pressures surrounding the post-war national rural settlement have begun to have a considerable impact on rural policy. As we saw earlier, agricultural policy, under budgetary and other pressures, has begun to move in a more regionalised direction. Moreover, the advent of the urban–rural shift of industry and counterurbanisation have placed pressure on planning policy. It should therefore be of no surprise that a process of spatial differentiation has also become evident in national rural policy, notably in the Rural White Papers published in the mid-1990s by John Major's government (DoE and MAFF, 1995; Scottish Office, 1995; Welsh Office, 1996), and in the recent Rural White Paper for England produced by Tony Blair's Labour government (DETR and MAFF, 2000).

In Hodge's view (1996: 331), the White Papers produced in the mid-1990s by the Conservatives represented 'the most significant public assessment of rural policy' since the Scott Report. Interestingly, they were produced on a country, rather than a UK-wide, basis. This led each to reflect the particular concerns of the diverse national contexts. For instance, in the Scottish White Paper, *Rural Scotland*, the primary aim was to ensure that 'Scotland's identity as a nation is enhanced', while in the English version, *Rural England*, it was to 'conserve the character of the countryside'. As Lowe points out, the differing concerns were reflected in the different policy goals adopted in each paper:

> The Scottish and Welsh documents are preoccupied with the question of how to sustain rural communities – how to ensure their cultural and economic vitality – whereas the English document is preoccupied with sustaining the countryside as a 'a national asset', reconciling economic and environmental objectives and ensuring that the rural environment and way of life are not 'submerged in our predominantly urban culture'.
>
> (1997: 390)

Within these differences in emphasis, all three documents stressed the diversity of the contemporary countryside. The English and Scottish papers were preceded by widespread public consultation exercises where the overriding concern expressed by most respondents was the local character of rural environments. The English White Paper took this recognition as its starting point and went on to assert that there is only so much that Government can do to foster local character and diversity:

> No single shift of policy, no universal scheme could provide for all this variety. The reality of life in the countryside is that many small-scale changes which respect the real differences in local circumstances are what are most likely to succeed.
>
> (DoE and MAFF. 1995: 6)

A recognition of diversity led to an emphasis on rural people 'helping themselves'. As the English paper put it:

> Self-help and independence are traditional strengths of rural communities. People in the countryside have always needed to take responsibility for looking after themselves and each other. They do not expect the Government to solve all their problems for them and they know that it is they who are generally best placed to identify their own needs and the solutions to them. In any case, local decision making is likely to be more responsive to local circumstances than uniform plans. Improving the quality of life in the countryside starts with local people and local initiative.
>
> (DoE and MAFF, 1995: 16)

All the White Papers gave considerable attention to the 'local'. Communities, partnerships and other local groupings should be incorporated into policy mechanisms because they so readily reflect diversity in the countryside. The job of rural policy is to somehow harness this local variety in line with broader policy goals.

While the White Papers produced by John Major's government were quite ready to recognise diversity at the *local* level there was some hesitation about invoking diversity between *regions* (notwithstanding the recognition that England, Scotland and Wales needed different policy formulations). However, with the election of a Labour government in 1997 a renewed emphasis on the regional level quickly became evident as one of the new Government's first acts was to allow referenda in Scotland and Wales on the introduction of a Scottish Parliament and Welsh Assembly. Following successful devolution campaigns, these two bodies came into being in 1999, and they now carry an enhanced responsibility for rural policy. In addition, the English regions were given more responsibility in the formulation and delivery of rural policy. MAFF (which has subsequently been renamed as the Department for Environment, Food and Rural Affairs) became drawn more closely into the work of the Government Offices for the Regions.[9]

Probably the most significant initiative in the English regions launched by the first Blair Government was the Regional Development Agencies (RDAs). These bodies (which were modelled on the Scottish and Welsh development agencies) were given a responsibility for rural economic development. Of particular importance was the fact that the RDAs acquired some of the functions of the Rural Development Commission, including responsibility for regeneration in disadvantaged rural areas. In this respect, they were charged with: tackling economic, social and environmental issues in an integrated fashion; encouraging integration between different economic and social sectors; co-ordinating the work of public agencies with a view to tackling rural problems; and delivering a range of services to rural areas in ways that meet the needs of recipients (see Ward *et al.*, 2001).

The regionalisation approach was also evident in New Labour's own Rural White Paper (for England) published in 2000 (DETR and MAFF, 2000). In its introduction to the various measures, the White Paper acknowledged the degree of regional diversity now existing in the countryside. It says:

> It is dangerous to generalise about the countryside, since different areas face different problems. The things that concern those who live in Alton, a market town in Hampshire, will be very different from those of concern in Alston, a market town in Cumbria. The contribution of land-based businesses to the economy of the Home Counties is limited and farmers struggle to manage land in the green belts around major cities, where their holdings are fragmented by urban service infrastructure and subject to vandalism. Remote counties such as Shropshire, Lincolnshire and Cumbria suffer more from declining farm incomes. In the south west and along the east coast fishing remains a vital industry. In the remoter areas the local and health authorities find it hard to provide services, such as meals-on-wheels and clinics. Other areas of countryside are suffering from structural change due to coalfield closures. As the Forest of Dean has showed, this type of change can take a generation to tackle. Counties like Durham, Nottinghamshire and South Yorkshire face considerable physical and community decline.
>
> (DETR and MAFF, 2000: 10)

In seeking to address this regional diversity, the White Paper put a great deal of emphasis on the activities of the RDAs, notably in boosting economic development in deprived rural areas.

The Labour Government also began to place a new emphasis upon regionalisation in planning policy. Traditionally regional planning has played a secondary role to national policy frameworks interpreted and administered in local (county and district) contexts (Baker, 1998). The first Blair Government quickly made it clear it was keen to upgrade regional planning from its previous advisory role. In 1998 the Government published a consultation paper entitled the *Future of regional planning guidance*. The paper began by stating that 'the interests of the English regions have been neglected in recent years and this government intends to reverse that neglect'. It proposed to do this 'by improving the arrangements for co-ordination of land use, transport and economic development planning at the regional level' (DETR 1998a: 2).

These new arrangements were spelled out in Planning Policy Guidance Note (PPG) 11, which deals with regional planning, and was first published in draft form in 1999 and in final form in 2000. This guidance note suggested that regional planning should now do more than simply echo national planning policy. Rather, it:

> places greater responsibility on regional planning bodies, working with Government Offices and regional stakeholders, to resolve planning issues at the regional level through the production of draft Regional Planning Guidance. This will promote greater local ownership of regional policies and increased commitment to their implementation through the statutory planning process.
>
> (DETR, 2000b: 3)

PPG 11 set out the new arrangements for regional plan preparation and emphasised the potential for policy development at the regional level. It proposed that regional plans should provide a concise spatial or physical development strategy, take advantage of the range of development options that exist at that level and be specific to the region. While the regional plan policies should refer to national policies, they should 'not simply repeat them nor resort to platitudes' (ibid.).

Thus, rural planning is now set in a stronger regional context. The regional planning authorities must set out strategies to guide local planning authorities in their administration of local rural areas. This approach raises the possibilities of differing planning policies tailored to the requirements of differing regional circumstances. A stronger regional tier could enable, on the one hand, much greater sensitivity to varied rural circumstances and, on the other hand, better recognition of the interdependencies between rural and urban areas.

These policy initiatives clearly acknowledge regional and local diversity in rural areas and imply that the challenge for government is to find policies that meet the needs of the 'differentiated' countryside and to develop the right mechanisms for delivering such a policy. No longer is a national rural settlement possible in the way it was after the Second World War; there is now a need for a more complex array of initiatives and policy structures. While policy has only just begun to shift in the direction of such complexity – many of the planks of the post-war settlement, including the CAP, remain in place – it is nevertheless possible to find clear evidence of differentiation in the rural policy process. The implications of this regionalisation of the policy framework are assessed in Chapter 7.

Conclusion

We have seen in this chapter that the countryside has changed in meaning and significance over the course of the post-war period. The initial tone was set by the strategic concern for increased food production during the Second World War. At that time rural land was required to meet the need for food, while the rural economy was seen primarily as an agricultural economy. These assumptions were carried through – largely via the influence of the Scott Report – into post-war legislation, notably the 1947 Agriculture Acts and the 1947 Town and Country Planning Act. These two acts were based on a simple belief – agricultural land should be protected from urban development in order to maximise food production from domestic sources.

However, since that time, this objective has been undermined for three main reasons. First, agricultural development has protected neither the rural environment nor the rural economy; it has led to extensive environmental damage and has entailed a continuous loss of farms, farmers and farm labourers. Second, new economic uses of rural space have emerged, associated with mobile manufacturing and service industries. In the process, the sectoral composition of many local rural economies has come to resemble that of urban economies, but with differences between extremely dynamic economic regions (e.g. the South East of England) and lagging regions (e.g. the North East of England). The industrial differences between these regions tend to outweigh any similarities between, say, rural Northumberland and rural Kent. Third, new social uses of rural space have emerged so that protected village locations and other attractive rural areas have become very desirable residential environments: house prices have risen, the social structure has become more middle class and environmental politics has become more central to the rural polity. Again, however, such social uses vary between rural areas – for instance, regions such as the North East of England display much less evidence of counterurbanisation than regions in the south of England. One consequence is that environmental activism is unevenly distributed across the countryside (Lowe *et al*., 2001).

These regionalisation processes have also promoted a regionalisation of policy. The recent Rural White Papers indicate that national rural policy is now much less significant than it has been for much of the post-war period. Instead, a greater regionalisation – both among the various countries of the UK and the regions within those countries – has taken place alongside a growing localisation – an emphasis on communities 'helping themselves' and taking more responsibility for the delivery of local services and other social benefits.

We can therefore argue that the structural context in which processes of rural change play themselves out has begun to shift. The economic, social and political formations that have worked to provide a strong national regulatory framework around development have begun to give way to formations that are increasingly orchestrated at the regional and local levels. And, as we shall see in subsequent chapters, these regional and local formations rely upon network configurations in which varied sets of conventions are asserted. The combinations of networks and conventions that comprise the more complex processes of change that currently run through rural areas effectively demarcate contrasting 'environments of action' in the countryside and ensure that differing rural areas are placed on differing trajectories of development. It is hard to see how these divergent rural regions could be incorporated into the kind of national policy framework that prevailed for most of the post-war era. Differentiation in policy will inevitably follow differentiation in development.

3 Theorising differentiation

Introduction

The struggle to re-define national policy in the light of complex changes in the nature of rural space finds its counterpart in academic discussions of 'rurality'. As long as agriculture reigned supreme in rural areas, it was perhaps appropriate for academic work to focus upon this sector as the mainstay of rural distinctiveness. Consequently, a great deal of analytical attention was placed upon modes of political representation and upon the economic structure of the industry (Newby, 1981). However, with the emergence of rural economies and rural societies increasingly detached from the agricultural sector has come a shift towards more general and holistic frameworks of analysis. These frameworks have been drawn from a range of sources, including political economy (Cloke, 1989), class analysis (Hoggart, 1997; Murdoch, 1995b; Phillips, 1998), state theory (Cloke and Little, 1990; Goodwin, 1998), and post-structuralism (Lawrence, 1997; Murdoch and Pratt, 1993). Despite their different emphases, these various approaches have all focused analytical attention upon the relationships between urban and rural spaces, between economic and social processes, and between global and local actions. As a consequence, academic theorising has been forced to seek out new, non-agricultural definitions of the 'rural' (see Hoggart, 1990; Newby, 1979).

Another main effect of the shift in theoretical orientation has been recognition of rural *diversity*. Rural development during the post-war period seemed to be heading towards greater homogenisation. However, the (re)discovery of *uneven* rural development, linked to the incorporation of rural areas into wider processes of economic and social change, has undermined taken-for-granted assumptions about a national rural structure. The nature of contemporary changes also indicates that not only is the rural economy complex in nature but that its make-up owes a great deal to forces emanating from *outside* the rural sphere (Cloke and Goodwin, 1992). The division between the rural and the non-rural thus begins to break down (Murdoch and Pratt, 1993). At the same time, the impact of general economic and social trends is seen to vary across rural space as such trends interact with local conditions (Bradley and Lowe, 1984). Thus, geographical diversity emerges more strongly on the academic agenda and this further undermines the scope for a new, all-encompassing definition of rurality (Pratt, 1996).

As we have emphasised in the previous two chapters, we need to consider the forces of differentiation and the way rural spaces become marked off from one another – economically, socially and politically – if we are to develop any understanding of the contemporary countryside. In this chapter we therefore examine some of the theoretical resources that have recently been employed to gain such an understanding. The account follows the structure of the previous chapter so that we first examine how the post-war settlement has been understood within the political science literature. We concentrate on analyses of the political structures that have underpinned this settlement and indicate that these analyses have recently shifted from the investigation of national state structures to a new concern for multi-level policy networks. Political analysis of the rural thus implicitly, and sometimes explicitly, recognises the emergence of rural differentiation and such differentiation is now thought to be underpinned by a reformed policy structure.

We then turn to document the same kind of theoretical shift in the economic arena. We argue that growing complexity in the rural economy – notably the growth in its non-agricultural components – has encouraged scholars to link rural economic processes into more general economic trends. In so doing, they have employed a variety of theoretical frameworks. We compare two main approaches: the *political economy framework* and *network analysis*. We consider how these two theoretical perspectives link rural economic activities to broader patterns of economic change and show how 'networked' rural economies are becoming differentiated from one another.

Turning, thirdly, to the social sphere we find the same type of theoretical resources being utilised in the wake of counterurbanisation. Where rural communities were once seen as having a holistic (almost structural) integrity, they are now thought to emerge out of the confluence of interacting social groups and social processes. The interaction between differing groups in rural communities undermines any notion that the community has a structural coherence over and above the various relationships that run through it and around it. These relationships – which are becoming increasingly complex as more and more groups attempt to utilise rural space – can be thought of as 'networks'. As we outline below, the new networks operating within rural communities lead to the emergence of 'reflexive' social entities in rural areas, that is, social forms oriented to the protection of the countryside's aesthetic qualities. Increasingly, we argue, the practice of aesthetic reflexivity leads to the assertion of a particular mix of conventions, one that prioritises environmental and localistic values.

In sum, we argue here that differentiation in the countryside is underpinned by the practices of political, economic and social networks. These network practices vary from network to network but in general we can propose that political networks operate according to strategic and bureaucratic standards, economic networks act in line with instrumental modes of calculation, and social networks conform to principles of cooperation and sociability. As we indicate below, each network type pursues particular mixture of conventions, in line with these various standards, principles and calculative modes.

The main objective of this chapter, however, is to assess academic perspectives on the differentiated countryside. Thus, we indicate that the identification of new political, economic and social forms in the countryside stems as much from new modes of theorising as it does from changes 'on the ground'. It is the *combination* of empirical trends *and* shifts in theoretical perspective that highlights the significance of new networks and conventions in the countryside. While we readily admit that our characterisation of these changes and shifts simplifies a number of complex debates and trends, we believe it has the virtue of allowing a broad range of issues to be covered within a single theoretical narrative. This theoretical narrative also provides a conceptual 'backdrop' to the case study chapters which follow in Chapters 4, 5 and 6.

Accounting for political change

We showed during the last chapter that for much of the post-war period rurality was closely linked to agriculture and that this linkage was made explicit in state policy where *rural* policy largely corresponded to *agricultural* policy. State policy was also dominated by *national* policy concerns so that *local* and *regional* differences tended to be subsumed into national modes of agricultural regulation and their associated conventions. It is therefore not surprising that state policy, and the institutional frameworks surrounding that policy, were the focus of much academic research. We begin by briefly reviewing the way scholars have understood the politics of the post-war settlement, in particular the stable policy structures that underpinned national agricultural policy, in order to show how changes in these structures have resulted in the emergence of new political networks in the countryside.

The political structures that dominated the agricultural sector emerged from the Second World War requirement to increase domestic food production rapidly. As we noted in the last chapter, post-war food shortages led the state into a so-called 'productivist regime', one which entailed resources being channelled towards increased output and productivity (Ilbery and Bowler, 1998; Wilson, 2001). The term 'productivist regime' refers to the network of institutions – state agencies, farming unions, input suppliers, financial bodies, R&D centres – that were oriented to achieving a sustained growth in food production from domestic resources (Marsden *et al.*, 1993). This regime dominated the agricultural sector in the second half of the twentieth century (and its legacy remains with us in the shape of the CAP). The overriding concern of the various institutions incorporated into the regime was 'modernisation' of the 'national farm' (Tracey, 1982).

Two regime institutions have been of particular interest to analysts of the post-war settlement: the MAFF and the NFU. These two organisations worked closely together in order to orchestrate post-war agricultural policy in line with the conventions of productivism. They evolved a stable and enduring policy-making relationship, one that effectively governed the entire national rural space (Self and Storing, 1962). Yet, while all analysts agree the MAFF/NFU relationship provided the central plank in the post-war policy framework, there is some dispute over its precise theoretical (or 'structural') status. In particular, there is disagreement as to whether this relationship is adequately captured by the term 'corporatism' or whether it is more accurately described as a 'policy network'. It is worth briefly considering this disagreement for it illustrates how the composition of political networks is implicated in broader structural shifts.

According to Grant (1983), the MAFF/NFU relationship is the 'classic case' of corporatism, with the Annual Review linking the two main actors together within a strong functional relationship. Cox *et al.* (1987) concur with this view and argue that such a policy structure was enshrined in the 1947 Agriculture Act which provided the foundation for corporatist arrangements in agriculture. Under the Act, the NFU was allowed to actively engage in policy formulation with the national state. In return, the Union would discipline its members so that they acted in accordance with the agreed policy.

However, Martin Smith (1989) questions this characterisation of the MAFF/NFU relationship. He believes that it was not as structurally robust as Grant and others have argued and he therefore re-assesses its 'corporatist' status. Smith says that farmers (in the form of the NFU) were involved in policy-making to only a limited degree: in his view 'the government ultimately decided policy' (ibid.: 84). Likewise, he claims the NFU played no major role in policy implementation. Smith believes that the relationship between the NFU and MAFF is therefore better described in policy network terms as a 'relatively closed policy community'. He says,

> It is clear that the agricultural policy community is very restricted and in the area of agricultural prices only includes the NFU. It is isolated from other networks, like the environmental network, and from the public and from Parliament. It has a shared view of agricultural policy and excludes groups which question the dominant agricultural agenda by not legitimising their access. It is a policy community which has corporatist features but it is not corporatist.
>
> (ibid.: 84)

In Smith's view the relationship should not be described as 'corporatist' because farmers were privileged but *junior* partners of the central state (see also Self and Storing, 1962). In this network of relations it was the state that defined national agricultural policy very much in line with its own objectives (namely, secure and inexpensive food through increased domestic food production).

Reviewing the corporatism/policy network debate, Michael Winter (1996) believes that, though the categorisation of the MAFF/NFU relationship as 'corporatist' remains a contentious issue, all interested commentators can agree that the agricultural policy community was marked out from other political networks by a number of special features. First, the agricultural policy community's participants shared a deep interest in maintaining the economic success of agriculture (the state for political reasons, the NFU for economic reasons). Thus, both groups emphasised production over consumption concerns. In this sense,

MAFF and the NFU constituted a 'community of common interest' (Winter, 1996: 102). Second, both groups shared the belief that increased resources should be channelled into the sector to ensure continued economic buoyancy. As Winter puts it,

> a policy community came into existence with a shared set of priorities and, particularly, an agreed need for public expenditure. In these circumstances, MAFF was able to harness the NFU to assist its case for lobbying the Treasury and other Cabinet ministers.
>
> (ibid.)

Third, the participants shared a common appreciation of the requisite priorities and how these should be met: 'The primacy of food production as a central objective for the nation as a whole provided in the immediate post-war period, a firm and clear focus for all members of the policy community' (Winter, 1996: 103). Fourth, the shared appreciation of priorities was underpinned by shared ideas, languages and underlying values. In Winter's view, this 'common culture helped to determine [the policy community's] internal success, its cohesion and its closure to other interests' (ibid.). Fifth, the cohesion of the agricultural policy community allowed it to remain relatively stable 'with a continuity of membership and boundaries which are recognised by the members of the community' (ibid.).

Winter sees these five shared concerns (and their associated conventions) as lying at the heart of the MAFF/NFU relationship. They allowed the agricultural policy network to maintain a remarkable amount of coherence for much of the post-war period (a coherence that was bolstered by the NFU's management of its membership, so that government always received a clear set of demands from the farming lobby). As a result, Winter (ibid.: 105–6) says, 'the MAFF–NFU axis became so strong that relations between ministry civil servants and NFU officers could be closer than between the civil servants and government ministers' (or, we might add, between NFU officers and farmers).

It is clear from these observations that a robust policy structure comprising MAFF and the NFU underpinned the post-war settlement. The two main participants agreed on the priorities for the industry and on how these should be implemented. Although there were frequent disagreements between the two groups, these were never severe enough to completely undermine established modes of working or dominant institutional structures. And the priorities agreed between MAFF and the NFU worked their way down through the policy hierarchy onto farms and into rural areas so that gradually rural England came to be re-shaped in line with the precepts of productivism. The everyday practices of farmers were increasingly 'disciplined' by the policy network in which they were enmeshed (Murdoch and Ward, 1997). Rural environments, communities and economies were subject to a productivist discipline, in which the economic conventions of efficiency and productivity were firmly to the fore.

The MAFF/NFU relationship provided a stable and robust structure of policy making and policy delivery. At the same time, agricultural policy was separated from the other policy sectors that bear upon rural areas. However, as we discussed in Chapter 2, this separation began to break down in the later years of the twentieth century. What undermined the division between agricultural policy and other rural policies was not only the increasing complexity of the forces shaping rural areas but also pressures emanating from within the agricultural policy community itself. Following the UK's entrance into the European Community, and its adoption of the CAP, the Annual Review lost its centrality in the agricultural polity as decisions over funding for the industry were now made in Brussels by the European Commission and the Council of Ministers (with the adoption of the CAP, agricultural policy was subject, for the first time, to a form of 'multi-level governance' – see below). Moreover, as we observed in the last chapter, entrance into Europe was soon followed by a period when the main agricultural commodities were effectively being overproduced. Surpluses and budgetary pressures therefore ensued. In the resulting agricultural crisis, the agricultural policy community failed to hold the line against competing political interests.

In accounting for the decline of agricultural dominance two main sets of changes can be identified. First, as we explained in the previous chapter, agricultural policy has been undone by its own success. Increases in output, production intensity and capital investment have revolutionised the agricultural industry but have also given rise to environmental, budgetary, trade and food quality problems. These problems have combined, leading to concerted pressure for policy reform both from within the policy community and from without. According to Winter (1996: 159), the agricultural policy community 'found that the depth of the crisis . . . rendered invalid traditional solutions and approaches'. Second, other policy relationships have emerged in rural areas. This is particularly evident in EU initiatives. For example, under the European Structural Funds, new ways of making decisions about rural development have been introduced in which rural local authorities and communities have gained a greater role in the design and implementation of development schemes (Ray, 1998; Ward and McNicholas, 1998). Likewise, agri-environmental programmes have brought with them new consultative arrangements, involving conservation agencies and groups, at the EU, national and regional levels (Whitby, 1996). All these approaches circumvent the top-down governmental approach associated with the traditional agricultural policy community and give rise to new priorities for action in the countryside.

In short, a governmental process of 're-scaling' (MacLeod and Goodwin, 1999) seems to be underway in the rural policy arena. National policy institutions (such as the Annual Review) have lost their dominance to a much more diffuse network of agencies distributed across various spatial scales (European, national, regional, local) and across sectors (agriculture, economic development, planning, environment). The rural policy framework is now best characterised as a 'multi-level governance' structure of policy delivery (Goodwin, 1998), one held together by a diverse range of political conventions. In this new policy structure, political networks emerge that encompass a much broader range of social and economic interests than was the case during the period of agricultural productivism. Under the post-war settlement, the state built strong linkages to farmers and their representatives but neglected other rural groups and organisations. In the more recent period of rural 'governance' the state finds itself orchestrating economic and social actions in a diverse range of fields, fields moreover that increasingly intersect. In short, rural regulation has become both more diffuse and more complex.

This broad shift in the nature of governmental relationships also has spatial implications. Edwards and colleagues (2001) argue that the use of governance mechanisms – such as partnerships – is leading to both a regionalisation (the RDAs, structural funds) and localisation (community initiatives, parish councils, LEADER) of rural policy. Edwards *et al.* say,

> the opportunity and capacity for partnership formation and operation are frequently geographically contingent, depending on funds for regional eligibility and/or the presence of institutions and individuals proactive in partnership promotion such as development agencies . . . and on the local presence of organisations in the voluntary sectors with the resource capacity to participate. These factors have a cumulative effect such that some areas emerge as 'partnership rich', reaping the benefits of additional funds, and other as 'partnership poor'.
>
> (2001: 307)

Thus, spatial and policy differentiation appear to reinforce one another within rural areas as particular policy structures interact with given socio-spatial formations.

We should note, however, that while the agricultural and rural development arenas appear to be moving into the 'multi-level governance' mode, not all parts of the rural polity are moving in the same direction at the same speed. For instance, in the land-use planning sector the situation is not quite so clear-cut. Here local communities and interest groups have long been involved in a local planning process that has largely been subject to local democratic control (Rydin, 1999). Rural communities and interests have been particularly influential in participation processes and this influence has increased as local plans have incorporated rural areas in the wake of the 1991 Planning and Compensation Act (Abram *et al.*, 1996). Thus, we might conclude that planning has traditionally displayed aspects of 'governance' (see Vigar *et al.*, 2000). However, recent attempts to open up planning to greater participation at the local level have been accompanied by a strengthening of centralised direction, most notably through the Policy Guidance notes issued by central government. The notes send policy principles down to the local level and these principles must be incorporated into local plans. Local planners are therefore 'disciplined' by the PPGs and local policies frequently 'mirror' central policies (Allmendinger and Tewdwr-Jones, 1997).[1] Thus, while multi-level governance in the agricultural sector means that national (or even European) policy is beginning to be re-defined as a more spatially sensitive approach, in the planning sector it implies that locally-sensitive initiatives are becoming coralled into a national policy framework (Murdoch *et al.*, 2000).

Such sectoral diversity in governance mechanisms indicates that in using a network perspective it may be necessary to place rural policy networks on a continuum, extending from policy sectors where governance mechanisms are weak to sectors where such mechanisms are strongly evident. As we have seen above, the MAFF/NFU relationship might be seen as a tightly-knit 'policy community' in which a shared understanding of policy goals is pursued by a restricted number of partners. These shared understandings include notions of productivity, efficiency, modernisation and so on. Rural development policy, on the other hand, would seem to be conducted primarily through associations and coalitions involving a large number of 'partners'. Such arrangements might be thought of as 'issue networks' in which competing and diverse aspirations for policy are negotiated (Smith, 2000) These competing aspirations include notions of welfare, community development, local capacity building and so on. Agri-environmental and planning policies appear to be intermediate cases that combine networks of both types; that is, they include quite a large number of participants but cover a restricted number of shared assumptions.

In these differing policy fields the actors enrolled in the new governance arrangements will find themselves subject to varied modes of negotiation, bargaining and control (Rhodes, 1997). Moreover, the differing networks will be bound together by differing political conventions. For instance, the agricultural policy community has, for much of the post-war period, promoted conventions of productive efficiency (often to the detriment of alternative conventions associated with market performance, local diversity or environmental benefit). The new rural development networks combine economic, localistic, civic and environmental conventions in line with the diverse objectives of their various members. In their different ways these networks define and shape the nature of rural space in line with their requisite convention hierarchies. Thus, any shift in the structure of governing institutions and in the composition of the networks can lead to a profound shift in the objectives of policy.

Accounting for economic change

As the previous chapter has indicated, the move away from an agriculturally-based rural policy was provoked not only by the crisis of agriculture but by a recognition that rural economies are now overwhelmingly dominated by non-agricultural manufacturing and service industries. In the light of these developments, analysts have come to realise that broad distinctions between urban and rural areas need re-thinking. In particular, it has been argued that such distinctions now need to be set within a *regional* framework. For instance, Saraceno (1993: 451), writing in the Italian context, claims that an urban–rural dichotomy explains very little of recent economic trends in the countryside for it was only during the pre-war and early post-war period that the geography of industrialisation conformed to the 'classical' spatial division of labour between the urban and the rural. However,

> more recently, with diffused industrialisation or other forms of decentral-

> isation, the rural/urban division of labour did not work in the classical pattern: different kinds of exchange – of labour, entrepreneurial capacities, capital, goods and services – between different sectors of activity, including agriculture, took place *in the same geographical space*, occasionally concentrating non-agricultural activities in the urban centres, but not necessarily doing so.
>
> (Saraceno, 1993: 464, emphasis in the original)

Saraceno (ibid.) thus argues that 'the "regional" logic . . . seems to explain much more than the "rural/urban" one'.

While rural development in Italy undoubtedly follows a course that is conditioned by that country's very particular socio-economic structure, we believe it is worth examining Saraceno's claim in the English context. In so doing, we again trace a shift in the economic structure of the countryside in order to tease out how rural space is now being differentiated in economic terms. In displaying this shift we first consider efforts to develop a general understanding of rural economic change within a political economy framework. We show that although this mode of theorising helpfully connects rural economic changes to shifts in the broader economy, it also tends to downplay any significance held by rural areas as general processes of industrial re-structuring play themselves out across international and national spaces. We then turn to examine economic regionalisation processes from the network perspective and indicate how a geography of rural economic networks might be brought into view.

As the previous chapter indicated, the study of the rural economy could, during the first half of the twentieth century, legitimately be seen as the study of the agricultural economy (for instance, the discipline of agricultural economics reflected the notion that the rural economy should be studied as a different entity to other spatial economies). However, changes since that time have rendered agriculture increasingly marginal in economic terms (except as a user of rural land) and have highlighted the significance of non-agricultural manufacturing and service industries in rural areas. In terms of its overall structure, the rural economy now looks much like the urban or the national economy. The rationale for a specifically 'rural economics' seems to be have been undermined by this broad shift in the main economic activities conducted in rural areas. As a consequence, the rural economy has recently been studied using *general*, rather than *rural*, theoretical tools.

One of the most generalised perspectives to be adopted is the political economy approach. It sees the rural economy through the lens of the various restructuring processes that unfold within capitalist industry. Put simply, and rather crudely, political economy considers the 'rural' to be nothing more than the outcome of the use of space by various 'fractions' of capital. Day *et al.* summarise this view when they argue that,

> In order to comprehend the most significant dynamic processes in contemporary 'rural' areas it is necessary to focus upon the interactions between

> various types of economic activity, themselves to be understood as outcomes of complex rounds of investment; it is the particular forms taken by these interactions which will determine the special character of given localities, defined in terms of their overlapping roles in a series of spatial divisions of labour.
>
> (1989: 229–30)

In this perspective, the rural itself has no distinctive properties – it is simply a 'bundle of resources' that can provide the basis for capital accumulation.

As Day *et al.* indicate, processes of capital accumulation can be thought of as 'rounds of investment': one 'round' is superimposed on others in regular bouts of re-structuring. Massey uses a now celebrated geological metaphor to capture the way a local economy is formed by the successive roles it performs in processes of accumulation:

> the structure of local economies can be seen as a product of the combination of 'layers', of the successive imposition over the years of new rounds of investment. . . . Spatial structures of different kinds can be viewed historically (and very schematically) as emerging in a succession in which each is superimposed upon, and combined with, the effects of the spatial structures which came before. . . . So if a local economy can be analysed as the historical product of the combination of layers of activity, those layers represent in turn the succession of roles the local economy has played within wider national and international structures.
>
> (1984: 117–18)

In this general approach, there is no longer anything distinctive about the rural economy – any local economy is simply the product of its industrial past and its contemporary role in the national or international economy. Even agriculture is regarded as a contingent industrial activity. Rees, for instance, proposes that the predominance of agricultural production in rural areas is simply the result of investment patterns driven by the requirements of capital:

> In other words, agricultural production and its associated spatial manifestations are nothing more than a particular instance of much more general trends within capitalist production and the uneven spatial development thereby engendered. Indeed, it is perhaps doubtful that agriculture has [any real] specificity (i.e. in terms of the significance of land) when compared with other branches of capitalist production such as coal-mining, iron and steel production and so forth.
>
> (1984: 33)

In short, the predominance of agriculture in rural localities simply reflects previous 'rounds' of investment and there is no necessary reason why future 'rounds' should conform to past patterns.

The political economy approach provides a robust account of economic change in rural areas, in particular the urban–rural shift of manufacturing and service industries discussed in Chapter 2. In Urry's (1995: 69) view this shift occurs because capital is becoming 'spatially indifferent'. Such indifference follows from an attempt on the part of capitalist industries to reduce their dependence on 'particular raw materials, markets, sources of energy, areas of the city, supplied of skilled labour and so forth' (ibid.). However, despite a general attempt to gain more autonomy in locational decision-making, Urry also argues that access to labour power (skills, cost, supply, organisation, reliability) becomes 'of heightened importance because it, unlike the physical means of production, cannot be produced capitalistically and hence is not subject to the same process of geographical levelling or homogenisation' (ibid.). In this regard, rural areas may gain new locational advantages as the shedding of labour from primary sectors means that there is a cheap, non-unionised, but flexible and technically competent workforce available for hire (Urry, 1984).

As this brief synopsis indicates, the explanatory strength of the political economy perspective resides in its ability to tie together a whole host of factors – including the requirements of capital, the re-structuring of industry and changing rural labour markets – within a single, holistic framework. Moreover, it allows rural areas to be considered in exactly the same terms as other spatial zones by focusing upon the underlying structure of the economy and the way this structure underpins particular uses of space. No specifically *rural* mode of analysis is required: urban and rural areas are bound together in *general* processes of economic change.

In explaining varied local outcomes in general analytical terms, political economy places considerable explanatory emphasis upon the requirements of capital as opposed to, say, the ability of local actors to modify processes of capital accumulation in line with their own interests or needs. This aspect of the theory means that it sees local changes as little more than the manifestations of broad, structural trends. For example, Urry (1984) claims that it is the supply of available labour that determines the attractiveness of particular local areas to capitalist enterprises. The same emphasis is evident in Massey's 'geological' approach, which has a tendency to see 'national processes in combination with, or embedded in, particular conditions producing the uniqueness of local economic and political structures' (Massey, 1984: 194–5). The production of local specificities is thus seen as a contingent outcome of a particular configuration of external forces (see Marsden *et al.*, 1993, esp. chapter 6; also Warde, 1985).

A more active role for local social formations is highlighted by Thrift (1987) in his consideration of linkages between the urban–rural shift of industry and counterurbanisation. In focusing upon the middle (or 'service') class, Thrift argues that this social formation does not just follow the location of industry but also plays a large part in generating a particular locational dynamic (such as the urban–rural shift) in the first place. Thrift gives a number of reasons for taking this view, including: the key role played by managers and professionals in determining economic location decisions; the need for employers to seek out pools of middle-class labour; and the way middle-class neighbourhoods attract particular services and thus more service sector employees. Thrift concludes that, 'as more service-class members arrive in an area so work and consumption opportunities, communications, and housing will follow' (ibid.: 79).

In this account Thrift shows how a complex interrelationship between the middle class and new patterns of economic activity helps to explain the growth of hi-tech businesses and professional services in accessible rural locations. Such growth can be seen to follow from the locational decisions of key personnel, drawn to rural areas by environmental and social characteristics (those very characteristics that underpin counterurbanisation). This aspect of economic change can also be seen to fuel the indigenous growth of small firms as new patterns of 'enterprising behaviour' become evident in rural locations (Keeble and Tyler, 1995: 994). Keeble and Nachum (2002: 74) argue that the renewal of 'enterprising behaviour' in the countryside can be ascribed to the growing numbers of mobile professionals that wish to migrate from congested metropolitan areas for 'quality of life' reasons. They say, 'these migrants bring with them know-how, expertise and client networks derived from their previous big city employment that enable entrepreneurship and successful new enterprise creation in their chosen small town or rural location' (ibid.).

The residents of rural localities may thus be more active in the shaping of economic space than is normally allowed for in the political economy framework. Rural economic changes should not be regarded therefore as simply the unfolding of general industrial structures, but as the interaction *between* local socio-economic formations and more general economic trends, an interaction that can be shaped as much by the local participants as by larger forces and structures. Moreover, the interaction between economic change and social change also implies that the rural qualities of particular localities may be important in accounting for the emergence of the new economic formations that reach across the urban–rural divide. It is the character of the rural environment, with its particular mix of social and natural attributes, that proves so attractive to business decision-makers. Economy and environment become closely intertwined in the rural context.

The significance of rural space (as a particular production environment) highlights the need to conceptualise international, national and local trends in a way that does not assume that one level necessarily dominates the others. There is thus a need to introduce a more 'symmetrical' approach, one that balances the various spatial scales of economic activity. One means of achieving such symmetry is arguably the network approach. In this perspective, the divisions between the international, the national and the local are re-conceptualised as 'network links'; that is, network associations are seen to transmit actions and decisions taken in one place to actions and decisions taken in another. Interest therefore shifts from 'rounds of investment' and 'industrial layers' to *network connections*.

The network approach has the potential to span the divide that opens up in the political economy approach between broad economic structures and their expression in particular localities. Economic networks are constructed by economic agents but are shaped by the general processes of re-structuring that sweep through capitalist economies (e.g. the shift from manufacturing to services). Any particular network can be seen as both a 'condensation' of broader structural processes as well as a re-configuration of these processes by active and knowledgeable actors. In other words, the network perspective implies a more 'equal' treatment of general economic processes and their local 'expressions' (Latour, 1993).

The rationale for adopting a network approach in the study of rural economies also arises from a widely held belief that contemporary processes of economic change are now generating network forms of organisation. For instance, the dynamic regional districts or clusters, mentioned in Chapter 1, are thought to be orchestrated by networks. As Staber observes:

> Networks are seen as an important defining characteristic of industrial districts, binding firms together into a coherent and innovative system of relational contracting, collaborative product development and multiplex organisational alliances. All economic action in industrial districts is said to be embedded in a dense web of network ties among individuals, firms and service organisations.
>
> (2001: 537)

In this account it is argued that firms, institutions and other economic actors are encouraged to 'cluster' together in network formations by the existence of 'agglomeration economies'. In general terms, such economies stem from four main characteristics of contemporary economic networks:

1 networks congeal around collective resources such as shared infrastructures;
2 the existence of industry networks in one place means that local labour markets come to hold specialist skills that can be utilised by firms and others;
3 firms can reduce their costs within spatially proximate interfirm transactions; and,
4 the clustering of economic actors facilitates knowledge transfer, innovation and learning.

(Malmberg and Maskell, 2002)

These characteristics provoke the formation of network clusters, places where 'mutual knowledge, collaboration and the exchange of information' are facilitated and where 'trust and mutual respect' are fostered (Maillat, 1996: 75). As is now well-known, regionalised networks have been identified in such diverse circumstances as northern and central Italy, Silicon Valley and Baden Wurttemburg (see, for instance, Cooke and Morgan, 1998).

The role played by new industrial network formations in generating economic change in rural areas, although under-researched in the literature, has been investigated in recent work by Keeble and Nachum (2002). In their study of clustering processes within the business and professional service sector in South-East England, they show that agglomeration economies do indeed span urban and rural locations. However, they suggest that such networks may be less significant in the rural realm than in metropolitan areas. Keeble and Nachum argue that, because rural firms tend to be established by *new* rural residents, they are (at least in the initial stages of their growth) likely to be less 'embedded' in the rural locale. The network linkages of the rural firms thus tend to be more dispersed than is the case for urban firms (rural firms have national, rather than regional or local, links).

Keeble and Nachum emphasise that the most important influences on rural service firms are the quality of the environment and transportation linkages. They conclude that,

> [our] survey shows that the overwhelming reason for choice of a decentralised [i.e. small town or rural] location was personal residential preference, evident in the dominance of 'proximity to founder's home' as the reason for location, and the exceptionally high and distinctive rating of 'attractive living environments' as important for the firm's competitive performance.
>
> (2002: 86)

In short, Keeble and Nachum argue that networks are more strongly clustered in urban than in rural locations, in part because so many business founders in the latter areas come from outside the region and the locality before setting up their businesses. Thus, agglomeration networks seem to be less significant in rural than in urban locations.

Yet, this finding should not be taken to imply that the new networks are unimportant in rural areas for Keeble and Nachum show that a significant proportion of rural businesses and professional service firms (57 per cent) do in fact *have* regionalised networks. While the figure is lower than for their urban (London) sample, it indicates that a significant number of rural firms are being incorporated into network formations established at the regional level.

Evidence gathered in the North East of England by Laschewski and colleagues (2002) supports this view. In their analysis of local business networks these authors show that efforts are being made by both state institutions, such as Training and Enterprise Councils, and more informal groupings, such as business clubs, to develop stronger network linkages in rural areas. A 'package' of networking activities has been introduced to the area including social events, circulation of bulletins, information services, workshops and focus groups. Interestingly, Lashewski *et al.* (ibid.: 383) discover that it is newcomers who are most active in the networks. They say: 'Newcomer businesses appear to see greater value in network formation, because of their own lack of contacts and support within a new locality and their experience from being located in other business contexts'. Thus, we might conclude that regional economic networks are emerging in the wake of counterurbanisation-led economic change in the countryside.

These findings allow us to make some general observations about the relationship between economic networks and rural areas. First, the very particular physical environment that marks out rural space now serves to attract economic actors that are seeking a good quality of life. Once established in a rural area, these actors may engage in 'enterprising behaviour', thereby strengthening both regional clusters and the reach of rural economic networks (Keeble and Nachum, 2002). Second, the productive activities that have historically taken place within rural areas can yield industrial structures that interact in very particular ways with the new industrial networks. For instance, Cooke and Morgan (1998) identify how small enterprises, which are often linked to traditional rural industries such as craft production and agriculture, seemingly enable network-like characteristics to be constructed on a local and regional basis (see also Asby and Midmore, 1996; Moseley, 2000). Third, the traditional character of the rural economy can inhibit the construction of new economic networks, especially in more remote locations. It is for, instance, recognised that such areas lag behind in the provision of technological networks (Talbot, 1997). Moreover, the narrow range of industrial activities in such areas may make the establishment of new network structures a difficult task (Murdoch, 2000).

This geography of economic networks also indicates that in differing rural areas we are likely to find differing mixtures of economic conventions. In areas marked by 'enterprising behaviour' we are likely to see conventions associated with industrial and market efficiency being combined with conventions of civic and ecological worth (although, as we shall see in following chapters, there can also be tensions between these convention types). In more traditional areas localistic conventions may be aligned with new market conventions (associated with innovation and trust) or equally may come into conflict with economic approaches. In more remote areas absence of economic growth and employment opportunities may result in greater attention to economic than localistic or environmental considerations.

In general, the shift to a network perspective entails a re-evaluation of economic resources and practices in rural areas. It shows that the performance of the rural economy is not to be assessed as simply an outcome of its greater incorporation into the general economic structure but stems from an alliance between local, rural economic processes and characteristics and more general economic trends. This alliance can be thought of in terms of network linkages and the rural economy can be conceptualised as a 'meeting place' for economic networks of various types and lengths (Massey, 1991: 23). Thus, 'success' in the rural economy might be seen as the development of vibrant networks that are able to best utilise economic resources in ways that maintain the social and environmental integrity of the countryside. However, we should note that the geography of rural economic networks interacts with the geography of social and political networks. As we shall see in our case study chapters, this interaction gives rise to differing combinations of conventions and patterns of development in rural regions and therefore different perceptions of 'success'.

Accounting for social change

The two shifts outlined above, in the political and economic frameworks that are thought to govern rural areas, highlight the importance of rural community involvement in development processes. Partnership mechanisms in the sphere of governance usually require community groups to take a more active role in the implementation of policy (Murdoch, 1997b). Rural economic networks also tend to rely upon processes of endogenous development in the generation of innovative strategies (Lowe *et al.*, 1998; van der Ploeg and van Dijk, 1995). These new mechanisms of rural development and change seem to imply a new role for the rural community.

Yet, in the wake of counterurbanisation, rural communities are apparently changing rapidly, with the consequence that their 'rural' characteristics are becoming enmeshed in 'urban' ways of life (Rogers, 1993). Now rural communities are thought to comprise diverse social practices and institutions and this diversity seemingly undermines the distinctive characteristics usually associated with rural places (Pahl, 1970). Thus, any integration of communities into economic and political structures must be predicated on the realisation that rural communities are no longer only 'rural'; they are made up of many differing forms of social life (Rapport, 1994). Moreover, fluid social relations are replacing the stable structures that had seemed coterminous with life in rural areas: 'people will be simultaneously participating in one "community", as a local network of interaction, whilst also being located in networks and "stretched out communities" of many other kinds' (Liepins, 2000: 30; see also Day, 1998; Silk, 1999; Stacey, 1969). In this section, we therefore follow the pattern established in the previous sections and analyse a general change in the character of rural communities by outlining how a shift in the social structures of rural areas is accompanied by the emergence of new networks.

The traditional view of rural communities is that they are consensual institutions in which kinship relations bind members into a coherent whole (Bell and Newby, 1972; Day and Fitton, 1975). A holistic view of rural communities has thus tended to dominate both academic analysis and the popular imagination (what Pahl, 1966, calls the 'village of the mind'). This view has been evident in the many rural community studies conducted in the post-war period where kinship networks, spatial proximity, close social co-operation and an agricultural orientation are thought to mark out distinctive rural social formations (see Harper, 1989 for a summary). In the wake of counterurbanisation, however, sociological interest has shifted from the role played by traditional social relations in maintaining the distinctive shape of rural communities towards the way general social processes play themselves out in the rural social context.

Pahl (1966), for instance, has claimed that in the light of social changes within rural communities, these entities are best interpreted not in terms of any particular spatial categorisation (such as urban/rural), but in terms of *social class*. He shows that the movement of middle-class commuters into the countryside has brought a more complex rural class structure into being. Where a traditional rural working class had previously seen itself only in opposition to a landed class, there is now an ascendant middle class.[2]

The work undertaken by Pahl in the 1960s was designed to show that rural social formations are increasingly becoming encompassed within a *national* class structure. He sought to establish that class relations in rural areas strongly resemble those in other spatial locations (see also Newby, 1979; Newby *et al.*, 1978). In other words, it is hard to differentiate the urban and the rural in terms of general sociological categories such as class for all spatial areas will be incorporated into general social structures.

However, despite Pahl's concern to undermine the significance of urban-rural divisions, his analysis implies that the interaction between the national class structure and local social formations retains a distinctively rural dimension. In part, this distinctiveness arises from the confluence of differing class groupings in rural locations, a confluence that is played out rather differently in rural than in urban locations.[3] Referring to class relations in villages, he says: 'Unlike, say, the suburbs, the village situation involves interaction [between the middle class and] other status groups' (1970: 274). The intensity of this interaction is what defines the character of villages and seems to mark off 'rural' from 'urban' or 'suburban' areas, for only in rural places are 'groups which, in the "normal" urban situation, would be socially distant . . . forced into an unusual consciousness of each other' (Pahl 1970: 275). Village communities thus retain something of their 'rural' character.

The maintenance of the 'rural' characteristics of communities in the countryside is also strengthened in certain respects by counterurbanisation. This paradoxical outcome is apparent in Bell's (1994) study of one village in the South East of England. In describing the social structure of the village, Bell shows that middle-class residence in the countryside is part of a search for new forms of belonging. In particular, middle-class residential preferences are strongly linked to the aspiration for a 'country' identity (to be realised through country sports, closeness to nature, village institutions and so on). In Bell's view, this linkage is made because rural residents believe a country identity somehow lies *outside* normal social relations (i.e. typical forms of class belonging). In the view of counterurbanisers (and, incidentally, traditional rural residents) the world of social interests resides primarily in the city, while the countryside operates according to a different, more 'natural' calculus.

In this account, counterurbanisation can be seen as an attempt to 'escape' the social through an immersion in 'country life'. Once established in rural locations, counterurbanisers will seek to consolidate those aspects of the rural that most closely accord with their pre-conceptions of this spatial area. This general attempt on the part of middle-class households to dissolve social tensions (such as class conflict) within rural communities can be seen, in Lash and Urry's (1994: 248) terms, as a form of 'aesthetic reflexivity'. Lash and Urry believe social reflexivity has become more pronounced in guiding social actions because of a general 'detraditionalisation' of social life. As Urry puts it,

> people's tastes, values and norms are increasingly less determined by 'societal' institutions such as education, family, culture, government, the law and so

> on. One effect of such stripping away of the centrality of such institutions is that individuals and groups are more able to envisage establishing their 'own' institutions, relatively separate from those of the wider society'.
>
> (1995: 220)

Urry proposes that the countryside has become a key site for the establishment of new institutions by the mobile middle class.

Counterurbanisation leads to the countryside being reflexively re-evaluated as a place in which 'authentic' forms of life can be established in order to offset the 'rootlessness' and 'placelessness' of much modern (that is, urban) existence.[4] Thus, the movement of middle-class households into the countryside can be seen as an attempt to create new relations, identities and forms of belonging following the breakdown of traditional social identities. The significance of the countryside, Lash and Urry (1994: 247) argue, stems not from processes of class formation *per se* but from a heightened aesthetic sensibility amongst middle-class groups. This sensibility is oriented to 'old places, crafts, houses, countryside and so on, so that almost everything old is thought to be valuable' (ibid.). For the reflexive middle class, country places are 'heavy with time' (ibid.: 250); they bear the markings of the past in their social and physical fabrics, and these markings are used to 'anchor' new middle-class identities.[5]

Counterurbanisation seemingly implies that social life in rural areas takes place in new social institutions and associations. While these institutions and associations might take a variety of forms, in Wittel's view they are likely to be established as networks. He provides the following rationale:

> 'Individualisation' presumes a removal of historically prescribed social forms and commitments, a loss of traditional security with respect to rituals, guiding norms and practical knowledge. Instead individuals must actively construct social bonds. They must make decisions and order preferences. [This is] a change from pre-given relationships to choice. Pre-given relationships are not a product of personal decisions; they represent the sociality of communities. In contrast, [individualisation] is defined by a higher degree of mobility, by translocal communications, by a high amount of social contacts, and by a subjective management of the network.
>
> (2001: 65)

Wittel describes this 'active construction' of social bonds by reflexive individuals as 'network sociality'. It comprises a social context in which the construction and maintenance of individualised (middle-class) identities depends upon the network linkages in which each individual is enmeshed. Wittel sees network sociality as antithetical to the traditional community:

> The term network sociality can be understood in contrast to 'community'. Community entails stability, coherence, embeddedness and belonging. It involves strong and lasting ties, proximity, and a common history or narrative

> of the collective. Network sociality stands counterposed to *Gemeinschaft*. It does not represent belonging but integration and disintegration. It is a disembedded intersubjectivity.
>
> (ibid.: 51)

If we take these arguments seriously, then counterurbanisation on the part of middle-class households indicates that traditional communities in the countryside are increasingly being displaced by networks in which individuals come together around shared interests. The nature of the new associations will be defined by the particular characteristics of these shared interests.

As we have seen, Lash and Urry (1994) believe counterurbanisers will utilise community networks to reflect upon the aesthetic values of country places, practices and environments. And as networks are consolidated, and as aesthetic aspirations are realised, the countryside is re-shaped according to a specific mix of localistic, communal and environmental criteria. This re-shaping is focused upon two main attributes of rural areas: communities and environments. As Blokland and Savage (2001: 224) indicate, social networks are processes through which 'agents mould spaces into places meaningful for their social identifications'. And networks are formed in local (rural) environments which allow people to 'get together and generate feelings of belonging – and hence of distinction – and communal identities'. And it is not only the community that is (re-)constructed within rural networks: according to Savage *et al.* (2001: 305), the establishment of network relations also requires some degree of control over 'external' relationships by which they mean 'the control (economic, social, cultural) of physical space'.

Our earlier work provided some evidence of this (aesthetic) re-evaluation of communities and environments in the countryside. In *Reconstituting rurality* we found that in the villages of Buckinghamshire, newcomers are extremely concerned to facilitate and generate communal activities. For new middle-class residents, rural life *means* life in a community, and if no such community exists, it will be (re-)created. In one of the case study villages we outlined how steady in-migration had resulted in a vibrant community emerging around thirty new village societies and community groups (see Murdoch and Marsden, 1994, Chapter 4). While (in Bauman's (2001) terms) this may not be the traditional rural community, it nevertheless serves to link rural residents to one another within a variable set of proximate social networks. Moreover, we also found that social networks were frequently invoked during environmental conflicts as counterurbanisers attempted to keep rural spaces free from unwelcome development. Processes of network formation thus require the simultaneous protection of both rural communities and rural environments. These aspects become bound into the 'defensive politics' of place practised by counterurbaniser networks (see also Barlow and Savage, 1986; Short *et al.*, 1986).

In sum, social change in the countryside leads to rural areas being systematically re-evaluated. The re-evaluation is undertaken by a middle class that, despite its urban roots, is imbued with elements of anti-urbanism. In developing anti-urban values the middle class falls back on pre-industrial ideologies related to the countryside and asserts these in a reflexive manner so that the characteristics of the countryside are consciously weighed against the characteristics of other (notably urban) places. The re-evaluation is also 'aesthetic' in the sense that traditional features of environment and community are seen as having heightened value. Such heightened value is asserted in a context where 'social life in the present is profoundly disappointing and . . . in important ways the past was preferable to the present – there really was a golden age' and its vestiges can be found in the countryside (Lash and Urry, 1994: 247). Given that this sense of disappointment is most pronounced in the urban arena, middle-class identities are naturally aligned with country identities (Bell, 1994).

The emergence of an 'aesthetic sensibility' among counterurbanising groups inevitably ensures that new social networks in rural areas will assert particular social conventions. In strongly counterurbanised areas, localistic and communal conventions will be consciously introduced. Ecological or environmental conventions will also be to the fore so that a new form of localism is established around the community and the environment. While, in many cases, these criteria may be drawn from rural traditions, their reflexive imposition demarcates counterurbanised rural spaces from those still dominated by traditional rural groups. In the latter places we can assume such traditions are implemented 'unreflexively'. Moreover, there may be strong support for the maintenance of genuine tradition in less counterurbanised communities (Day and Murdoch, 1993). In yet other places there may be something of a struggle taking place between proponents holding the two differing conceptions of rural space, one concerned for the aesthetics of rurality, the other concerned with the maintenance of traditional ways of life. As we shall outline in subsequent chapters, the assertion of reflexive and traditional conventions in the same rural space can provoke some amount of conflict, notably around levels of economic development.

Conclusion

The theories of political, economic and social change outlined in this chapter chart a shift from solid and limited national trends to diffuse and differentiated patterns of change. Such differentiation tends to be orchestrated by the networks that have come to the fore as the political, economic and social structures that have prevailed in the post-war period give way to more fluid social forms. In the political sphere the 'policy network' emerges as the new 'multi-level' state co-ordinates its various activities in line with new socio-economic demands. In the economic sphere we find that networks now lie at the heart of the new regional agglomerations. In the social sphere the growth of the 'reflexive' middle class promotes new networked forms of sociality within rural communities. The differentiated countryside is therefore underpinned by differentially composed and distributed networks. And these differing networks have varied goals and are subject to contrasting modes of evaluation.

The variations and contrasts in network, goals, objectives and modes of evaluation are especially important when we consider interactions *between* networks in the development process. A key aspect of this interaction will be the degree to which the prevailing hierarchy of conventions established in particular places by particular networks is pre-disposed to *support* or to *resist* development. For instance, we can speculate that in those areas where counterurbanisation is well-advanced, social networks may be employed to counter proposals for further economic development.[6] In more remote and traditional areas, the prevailing conventions may be strongly supportive of further economic development, but the 'thin' nature of the existing economic networks may make the generation of such development quite problematic. In both these cases we see an important interaction between social and economic networks, an interaction that is shaped by the attempted implementation of particular convention hierarchies in the social and material fabrics of particular rural spaces.

In the next three chapters we describe in some detail how the interaction between differing networks serves to shape patterns of development in the differentiated countryside. In Chapter 4 we examine the assertion of anti-development conventions in the context of planning in the 'preserved countryside'. We consider how counterurbanisers seek to exercise some territorial control in the buoyant economic context of Buckinghamshire and investigate whether 'preservationist' conventions can be maintained in an area where demand for development is almost unceasing. In Chapter 5 we look at processes of contestation between development and environmental interests in Devon, the 'contested countryside'. We show that in this location 'preservationist' conventions come to co-exist with more traditional, developmental aspirations so that the hierarchy of land uses established in the countryside reflects a compromise between the two. In Chapter 6 we turn to the 'paternalistic countryside' of Northumberland. Here we find weakly developed levels of counterurbanisation and much less preservationist sentiment. The rural area is still dominated by paternalistic landlords and their traditional local associations. Locally-embedded developmental conventions therefore define the nature of local land uses. In sum, these three case studies illustrate how economic, social and political networks act to differentiate rural regions and how the 'new' networks of middle-class counterurbanisers interact with more traditional rural socio-economic formations. This interaction, we argue, decisively determines the shape of the differentiated countryside.

4 The 'preserved countryside'

Introduction

In the previous chapter we identified a series of networks that act in both complementary and conflicting ways to shape rural areas. We argued that these networks emerge from consecutive processes of re-structuring in rural politics, economy and society and that they, and their associated conventions, underpin processes of differentiation in the countryside. In the following chapters we develop this analytical perspective and examine, first, how the networks and their requisite conventions are differentially distributed throughout the countryside and, second, some of the consequences that flow from this distribution. In so doing, we also consider the utility of the ideal types outlined in our earlier work and we consider whether the processes of regionalisation currently shaping the contemporary countryside can be adequately comprehended within our 'differentiated countryside' framework.

In this first case study chapter we examine the 'preserved countryside'. Our initial conception of this countryside type presented in *Reconstituting rurality* was that it can be seen most clearly in the South East of England where high numbers of middle-class activists use the planning system to press for strong control of development in ways that reflect their preservationist aspirations. In other words, the preserved countryside equates to a counterurbanised or middle-class countryside where the expression of 'aesthetic reflexivity' enshrines localistic and environmental conventions in the rural landscape.

As we mentioned in Chapter 1, other authors have challenged our characterisation of the preserved countryside. Hoggart, for instance, argues that there is far less evidence than might be imagined for this assumption of middle-class dominance in the preserved areas. He says that:

> The opposition of village residents to particular developments is very real but highly constrained. Better organised and more powerful players than the amorphous, service classes influence procedures, regulations and structures in the countryside. . . . It might well be that the service classes do not want growth in their locality but *ceteris paribus* there is little evidence that localised groups are capable of successfully resisting development when these are supported by powerful national interests.
>
> (1997: 259)

Hoggart appears to be suggesting here that to talk of a 'preserved' countryside is misplaced for preservationism can easily be undermined by the activities of powerful political and economic networks. In this chapter we investigate this issue in some detail. In particular we examine the interaction between political networks and social networks in the context of planning for housing. New housing is an especially contentious issue in preserved areas because it alters both the rural environment and the rural community. But because housing pressure is so strong, largely as a result of this countryside's attractiveness to counter-urbanisers, the threat remains almost permanently tangible. Moreover, because the planning-for-housing system is governed by strong national to local networks (Murdoch and Abram, 1998; 2002; Vigar *et al.*, 2000), a consideration of this policy area allows us to investigate how struggles for the control of territory take place between differing network types.

The context in which we examine the struggle around new housing is the county of Buckinghamshire in South-East England (the area studied in *Reconstituting rurality*). We first follow arguments around housing development within the review of the Buckinghamshire County Structure Plan, conducted during the early 1990s. We then turn to look at planning-for-housing issues at the more local level of Aylesbury Vale District Plan (which followed directly from Structure Plan policy). As various commentators have noted (e.g. Counsell, 1998; 1999), local plans are arenas in which a host of competing values associated with development, environment, social equity and so on, come to be asserted against one another. Analysis of the debates circulating around and through planning at the county and district levels helps to show how counterurbaniser networks affect development processes at the local scale, and how reflexive actions on the part of such networks impact upon planning policy.

In both the case study plans considered in this chapter the local preservationist coalitions are challenged by powerful national-to-local configurations of developmental actors, much in the manner suggested by Hoggart (1997). Local preservationist networks find themselves confronted by a developmental network that pushes economic conventions down from the national to the local level. As part of a response to this developmental configuration, alliances are established between local preservationist groups and national environmental networks in order to embed preservationist conventions in policies formulated at regional and national scales of governance. Thus, national-to-local networks can be identified in both the environmental and development arenas (see also Murdoch and Marsden, 1995).

In what follows, we identify changes in the spatial scale of preservationist politics by examining negotiations around Regional Planning Guidance (RPG) in the South East region. We show that the concerns of preservationist actors in local areas of the region were initially channelled into the South East Regional Planning Authority (SERPLAN) via councillors and other local representatives. SERPLAN's adoption of preservationist conventions was then bolstered by a campaign against the housing projections waged by the CPRE at the national scale. This campaign coincided with the review of South East RPG and inevitably shaped the ensuing regional planning policies. We therefore examine how a preservationist network extending from the local to the national scale was put in place and how this network was effectively able to challenge the national-to-local development network. In conclusion, we consider the character of recent regional planning policy and the likely future of preservationism in areas such as the South East of England.

The 'preserved countryside' in its regional context

The South East is the core economic region of the UK: London, a world city, not only casts a giant shadow over the area, but many of the leading economic and political institutions are also located there (Allen *et al.*, 1998). These characteristics usually ensure economic buoyancy and are reflected in the region's continued economic dominance of the rest of the UK – London and the South East account for around one third of the UK's GDP. The buoyancy of the region is also evident from its business structure, which is oriented towards the most dynamic sectors of the economy. The South East has a much lower proportion of its workforce in extractive and manufacturing industries than the UK as a whole and a much higher proportion in sectors such as distribution and finance. The service-based composition of the regional economy accounts in part for its success, although it does not imply that this economy is immune to recession, as the downturn in the late 1980s and early 1990s showed.

The high levels of employment and wages in the South East attract in-migrants from other regions of the UK and overseas (London is the main destination of new residents in the UK despite efforts by the Home Office to engineer 'dispersal' to other cities and regions). During the 1980s the South East region accounted for 58 per cent of national population growth, with the major part of this increase concentrated in the area around, rather than within, London. With a population of over 18 million the South East now has the highest regional population in the UK and is the second most densely populated region after the North West. Nevertheless, the population continues to grow: the Government's 1992-based household projections indicate a likely increase of 1.7 million households in the region between 1991 and 2016. These projections assume that there will be substantial international migration into London and substantial migration out of London into the rest of the region and beyond (SERPLAN, 1998).

Fielding (1992; 1998) has suggested that the movement of population in and out of the South East is structured along the lines of social class. In order to illustrate the class character of the migration process he introduces the notion of the South East as an 'escalator region'. Noting that, with its dynamic economy, the South East offers rich opportunities for social promotion and the creation of middle-class careers, Fielding argues that young middle-class adults will be drawn to the area from places lacking in such opportunities: 'young adults will tend to gravitate towards the former and away from the latter. Older people, some of them in-migrants as young adults, will tend to migrate away from the former towards the latter' (Fielding, 1998: 49). Thus, while young, well-educated workers tend to be willing and able to seek out good jobs in dynamic regions, by the time they reach middle age they often wish to move away from the most dynamic urban areas towards smaller towns and settlements, both within and beyond the South East. The region therefore acts as a 'social escalator', allowing young adults to move into the middle class while at the same time sending established middle-class members out into desirable residential areas in the suburbs and the shires (Fielding, 1998).

The combination of a buoyant economy and the operation of the 'social escalator' places a high status on living in the countryside and puts many rural areas – especially those located in the more economically prosperous parts of the region to the south, north and west of London – under a great deal of development pressure. It is no surprise, then, that the countryside in the prosperous sub-regions attracts developers. The demand for housing in such places is almost unrelenting, thus any housing development that does take place is assured of a ready market made up of those prosperous housing consumers that are riding the 'escalator'.

Situated in the north-western part of the South East region, the rural areas and small towns of Buckinghamshire are a key destination for many of those people moving out of Greater London and other cities. Population growth in the county has been particularly intense – between 1971 and 1991 Buckinghamshire had the fastest population increase in the UK, and rates have remained high since (Murdoch and Abram, 2002). The county has a long tradition of middle-class migration into the rural areas and its well-preserved southern districts have proved particularly attractive to wealthy in-migrants from towns and cities in South East England and beyond (Murdoch and Marsden, 1994).

While conducting research reported in *Reconstituting rurality* we discovered that villages in the county are becoming increasingly middle class in character. For instance, we found that, on average, around two-thirds of village residents had moved into their houses within the previous ten years and that almost half of all rural residents fall into socio-economic groups 1, 2 and 3, with a further third in the economically inactive (retired) category. Almost all the residents in these villages indicated that they were strongly opposed to further development, with the vast majority (70–80 per cent) wishing to see no more new homes or other development in the local countryside (see Murdoch and Marsden, 1994, chapter 2). Having stepped off the 'escalator' in the suburban and rural environments scattered across Buckinghamshire, the new residents soon showed themselves to be opposed to further development of such areas. Thus, following extensive counterurbanisation into the county, localistic and environmental conventions have come to dominate Buckinghamshire's countryside.

Reconstituting rurality proposed that a middle-class and preserved countryside is gradually coming into being in places such as rural Buckinghamshire. Partly through their dominance of rural institutions (such as local councils and community groups), and partly through their willingness to mount political campaigns against development, members of the middle class are gradually 'winning out' in the battle for rural space. Their aesthetic aspirations for the countryside come increasingly to be reflected in dominant patterns of land use. In the process, other claims on this space, whether from developers, working-class residents, urban youth or travelling people, are sidelined. Essentially, a cumulative process of class composition has been set in train (Savage *et al.*, 1993), for as preservationism succeeds in excluding 'other' forms of development so it excludes 'other' people.[1]

In what follows we examine whether this characterisation of the countryside in Buckinghamshire still holds. We focus on the planning arena as a domain where conventions of rural protection and preservation are likely to be reflexively expressed by middle-class networks. We also consider opposition to these networks from those actors that assert developmental repertoires of evaluation. We thus examine how the conflict between developmental networks and environmental networks shapes development and planning in the 'preserved countryside' and the extent to which preservationism can be maintained in this dynamic and fast-growing region.

The Buckinghamshire Structure Plan

Buckinghamshire is made up of five districts: Milton Keynes and Aylesbury Vale in the north; High Wycombe, South Buckinghamshire and Chiltern in the south. The initial growth centre was High Wycombe where, during the 1960s, the Labour-controlled council sanctioned a policy of substantial public housing investment. However, by the early 1970s a coalition of conservation groups, spearheaded by the local branch of the CPRE and the Chilterns and High Wycombe amenity societies, forced the adoption of a much more restrictive growth policy in the south of the county, an area which comprises mainly green belt and an Area of Outstanding Natural Beauty (AONB). It is evident that preservationist networks have long worked effectively within Buckinghamshire's planning processes. These networks have been active in promoting conventions associated with local and environmental qualities and they have ensured that each round of structure planning entrenches containment in the rural areas of the south and north of the county and concentrates development (much of it generated by London overspill) in the urban areas of the north (notably Aylesbury and the new city of Milton Keynes).

A general convention of containment has thus come to prevail in the county. Amenity societies, action groups, local residents, county and district councillors have evolved a common perception of the overall development context (i.e. a countryside and rural environment under threat from relentless urban economic growth) and have established associations that work to ensure no incursion of development into the rural areas in both the south and the north of the county takes place. The conventions of local protection and environmentalism have become bound into these associations. As Charlesworth and Cochrane put it:

> There appears to be a mutual understanding between the local planners and local residents of the need to retain strict planning policies and withstand

> pressures for development. Indeed, this can be seen as a mutually beneficial (and particularly powerful) alliance in which the traditional development control activities of planners are justified by the activity of local groups.
>
> (1994: 1732)

In short, local planners, politicians and residents assert preservationist aspirations and development plan review processes are the most obvious outlet for this assertion.

The most recent review of the Structure Plan began in the early 1990s and we briefly outline its chronology in what follows in order to show how the reflexive assertion of preservationist conventions is normally undertaken in Buckinghamshire and elsewhere. As we describe below, preservationism in planning is promoted by five main groups: political councillors, professional planners, developers, local residents and environmentalists. These groups frequently act in concert with one another in order to pursue shared objectives. Alliances between the groups are revealed in plan review processes (Murdoch *et al.*, 1999).

The Structure Plan review began with a group of planning officers putting together a series of papers on the contemporary state of the county, in terms of housing, employment and transport. The summary papers and survey results were then submitted to a Planning Panel, comprising ten county councillors. The Panel agreed an overall strategy for the new Structure Plan. According to a planning officer involved in this process, the political members were concerned to ensure that the south of the county – the green belt and the AONB – were preserved: 'they've always made clear right from the word go that they wanted the green belt protected and they wanted the AONB protected. You've got those two fixed starting points which they were firmly attached to'.[2] In other words, the Panel positioned itself as a key part of the preservationist coalition and asserted conventions of local protection and environment over those associated with market demand and economic growth.[3]

Yet, while the preservationist coalition could decide upon the *distribution* of development, it soon became apparent that it could not so easily determine the overall *level* of development (Murdoch and Abram, 2002). This was because the housing figures to be included in the plan cascade down to the local level from national housing projections. National government calculates future levels of housing demand and then disaggregates the total down to the regions, counties and districts (Vigar *et al.,* 2000). The purpose of the projections is to ensure that planning meets the government's objective of a 'decent home within reach of every family' (see House of Commons, 1998: 44). The role of planning is therefore to 'provide an adequate and continuous supply of housing' (according to DETR, 2000a: 2). In order for planning to fulfil this role, all tiers of the planning hierarchy need to work towards the same overall numbers and must allocate sufficient housing land to meet expected levels of demand. Thus, once a set of national projections has been derived, they are translated into regional and local housing requirements. In effect, the projections tie together the various tiers of planning within a network of 'demand-driven' projections (Murdoch and Abram, 2002).

During the Buckinghamshire Structure Plan review it became evident that the South East region (excluding London) would require around one million new houses to meet anticipated demand up to 2016. Buckinghamshire was expected to accept its share of this new housing. During the early 1990s the county figure was set at 62,600 dwellings for the plan period (1991–2011). A continuation of this level of development was likely to be acceptable to the County Council as it was keen to act as a reasonable and responsible member of the regional planning process. However, a debate about the implications of the figures was expected to shape the Structure Plan review. This debate would provoke an interaction between 'vertical' development networks, which attempt to push conventions associated with housing demand down to the local level, and 'horizontal' preservationist networks, which seek to promote the protection of rural environments.

To begin with, the preservationist coalition made all the running. It not only managed to get its protectionist principles enshrined in early policy statements but it also interpreted the government's emerging 'sustainable development' approach (as laid out, for instance, in PPG 12 published in 1992 – see DoE, 1992) to mean that new houses should be concentrated in existing settlements, near public transport access and near employment areas, rather than in areas of, say, open countryside. Thus, the policy of restraint that already existed in earlier rounds of structure planning received further legitimisation as 'sustainable development' became a central goal of planning policy (Murdoch and Abram, 2002).

Once this broad strategy was adopted, other policies naturally followed. The housing distribution figures allocated the vast bulk of new housing to the urban centres of Milton Keynes, Aylesbury and High Wycombe. On employment policy, the plan stated that most new economic development would take place in the north 'with a greater emphasis on restraint in the constrained south' (Buckinghamshire County Council, 1994: 41). Milton Keynes and Aylesbury were identified as the main employment growth centres. In the rural areas outside the green belt, 'small-scale employment-generating developments appropriate to . . . local needs . . . will normally be acceptable' (ibid.: 42). Following on from this, it was stated that 'the impact of new development on the Buckinghamshire countryside will be reduced as much as possible' (ibid.: 77). An increased priority was also attached to redeveloping urban land. Thus, the whole thrust of the plan was established in accordance with the overriding requirement to protect green belt and the rural areas. Conventions associated with environmental protection seemed to dominate those associated with economic growth and housing demand.

The main objections to this strategy came, unsurprisingly, from the house builders. Their opposition derived from a desire to open up more development potential in the lucrative south of the county, notably by moving some of the allocations given to urban areas in the north (e.g. Milton Keynes) down to the south (e.g. High Wycombe). The developers made their main challenge to the Structure Plan strategy during the Examination-in-Public (EiP).[4] As the Examination unfolded, it became clear, first, that if the house builders wished to increase the overall number of houses they would have to undermine the housing projections produced by the County Council; second, if they wished to alter the allocation of housing – shifting some substantial number down into the south of the county – they would have to argue against the green belt and AONB restrictions in the south. In order to do this they would have to reveal inconsistencies or unwarranted assumptions in the general strategy and the policies of the County Council.

Effectively the house builders attempted to open up the technical dimensions of the plan to political scrutiny. First, the argument was made that the plan needed to be less responsive to local political sensibilities and more concerned with market criteria so that further development could be permitted in areas of high demand (i.e. the south of the county). Second, the house builders took issue with the County Council's demand projections and attempted to get the numbers raised. The consequence of this second challenge was to ensure that the whole discussion about levels of development at the EiP was dominated by conventions of demand, with debates revolving around the County Council's defence of its own demand figures.[5] Because the EiP was dominated by the demand conventions (that is, the extent to which key assumptions within this discourse were reflected in the housing numbers), then (perhaps ironically, given the developers' desire to shift housing provision into the south of the county) the *distribution* of development received far less attention.[6] As a consequence, the conventions put in place by the preservationist coalition at the beginning of the review process remained largely unmoved.

These various arguments and conventions were combined in the final report of the EiP Panel. Using projection-based arguments the Panel raised the number of houses during the plan period to 66,500 but left the distribution largely untouched. Arguably, the house builders, through their deployment of technical arguments, had won a minor victory in terms of the small increase, although the County Council also felt its overall strategy had been vindicated. Yet, the preservationist coalition, established right at the beginning of the review process, had largely succeeded in withstanding the challenge of the development network so that in the finalised plan the majority of the houses were allocated to the largest urban centres of Milton Keynes, Aylesbury and High Wycombe. The rural areas in both the north and the south of the county were set within a general framework of protection.

The Aylesbury Vale District Plan

In this section we focus on the district of Aylesbury Vale where, following the Structure Plan review, the District Council set about formulating its own district-wide plan. Because the district lies just outside the green belt it has long been a target for development and, as we saw in the previous section, the town of Aylesbury, which has grown rapidly in recent years, was set to take a further round of new housing under the Structure Plan proposals. The problem of accommodating and distributing growth was always likely to be pronounced in the district. However, as we shall see, preservationist concerns were to the fore as the debate focused not on the town of Aylesbury (where the bulk of new development was to take place) but on the rural areas of the district.

Aylesbury Vale encompasses approximately half the geographical area of Buckinghamshire. The district lies broadly north of the Chilterns escarpment and extends from the town of Wendover in the south to Buckingham in the north, where it borders Milton Keynes. Aylesbury is the administrative centre of the district (and the county) and contains a population of almost 60,000, one third of the district total. As well as Aylesbury there are over 100 smaller settlements in the Vale. Development pressure is at its most intense in the south (close to the green belt), particularly around Aylesbury itself. The north of the district still retains a rural feel, thanks mainly to the policy of channelling development into the urban centres.

In general, the pace of social change in the district has been rapid. The small market town of Aylesbury was designated for expansion in order to accommodate London 'over-spill' from the 1960s onwards, and grew markedly in the 1970s, with a great deal of new building in and around the historic core. In the 1980s and 1990s, the population continued to grow, increasing by 20 per cent between 1980 and 2000, with Aylesbury receiving 8,000 new residents during the 1990s alone. The area has therefore been subject to intense development pressure.

The composition of the in-migrant population, especially that gravitating to the rural areas, tallies with that outlined earlier in the chapter – mostly middle-class families with children, looking for somewhere 'green and quiet' in this most prosperous and dynamic of regions, to set up home. And once these families have become resident in the district they tend to become concerned about threats to the quality of the community and environment. They thus tacitly or actively support those who get involved in planning processes in order to protect local areas from development (see Murdoch and Marsden, 1994).

Planning policy in the district has traditionally emphasised rural protection and the concentration of development in Aylesbury town. A *Rural areas local plan*, prepared in the early 1990s, for instance, stressed that policies would aim to 'protect and enhance the general environment and natural beauty of the country-side' (Aylesbury Vale District Council, 1991: 5). It advocated the continued development of Aylesbury and the protection of smaller rural settlements. A rather strict division between urban and rural areas thus appears to have been put in place in the Vale, mirroring the division in the county plan.

Following the 1991 Planning and Compensation Act and the review of the Structure Plan, the District Council began to put together its own district-wide plan. It started with the principle of 'sustainable development', interpreted in line with the urban containment approach, and put together a set of proposals that again specified that Aylesbury should be the main development node. However, a certain amount of disquiet within the Council, notably among political members, soon became apparent. This disquiet stemmed from a change in political control of the District Council during the plan review: initial growth rates had been agreed by a Conservative-ruled Council and a new Liberal Democrat ruling group displayed more concern about the housing figures emerging from the Structure

Plan (reviewed in the previous section). The new Council was faced with a dilemma: it could either make itself unpopular with constituents by seeking locations for large amounts of new housing or it could renege on commitments entered into with the regional planning authority.

This dilemma dominated the deliberations of the Strategic Plans and Development Committee. At one meeting, councillors made comments such as: 'it would be tragic if this county is to become semi-suburban sprawl'; 'we need to challenge some of the assumptions behind the figures'; 'we don't want more people to live in this area in considerable numbers. . . . Might we be so bold as to say that Aylesbury is full up?' At the end of this discussion it was proposed that Aylesbury Vale District Council tell the government that the district 'cannot cope' with further housing allocations. In response, the planning officers present at the meeting counselled caution, one saying 'it would be unfortunate if one of SERPLAN's [South East Regional Planning Forum] members says, we don't want (housing), look elsewhere'. The planning officer reminded the councillors that it was not good enough to say 'we can't take development'. In this brief exchange, the convention of demand was re-asserted by planning professionals keen to ensure that the Council remained a responsible participant of the planning-for-housing policy network.

In effect, the channelling of demand conventions down the planning network acted to bring the local politicians back into line. They were forced to abide by the housing figures that emerged from the Structure Plan. The local implementation of these demand conventions was undertaken by the planners who were less swayed by localistic and environmental concerns. In fact, the attitude of planners differed from that of the councillors in significant respects for some in the planning department seemed to see the new housing numbers as an opportunity to engage in creative planning, that is, planning for development rather than planning for preservation.

In Aylesbury Vale the planners pointed, for example, to opportunities to put into practice new theories about 'urban villages'. As one commented: 'There's a political will for Aylesbury Vale to kind of go places and to achieve more than it has done in the past and we've always accepted that encouraging growth into Aylesbury is a major part of that'. Within the planning department it was simply assumed that Aylesbury would continue to grow, in part because at that time, as another planner put it, 'there's certainly no big body of opinion which is anti-development in the town'. In other words, economic conventions easily coalesce with civic modes of evaluation in the beliefs and practices of the planning professionals.

The negotiations between the politicians and the planners (in the context of policy cascading down from above) shaped the early plan proposals.[7] The housing in Aylesbury was to be distributed in accordance with the principles of sustainable development, and was to be linked to the provision of a transport infrastructure, based on the 'urban village' concept. These 'villages' would be arranged along a new transport corridor. The district planners also employed sustainability criteria to select a number of key settlements for development in the rural areas. Rural housing was to be 'concentrated at a limited number of settlements which offer the best prospect for limiting the need to travel and, through offering a choice of transport, minimise the use of the car' (Aylesbury Vale District Council, 1998: 26). In the rural areas, the planners used sustainability criteria to direct development towards the largest villages with the best transport links. Around 6,200 new houses would be allocated to Aylesbury and just over 800 would be apportioned to rural areas.

However, as the plan review unfolded, local politicians and planners came under acute pressure from residents in the large villages to reduce the housing allocations. A proposal came forward that two of the growth villages should have their numbers reduced, with the remaining 400 houses distributed among a further 'tier' of smaller villages throughout the Vale. With a General Election looming the councillors were worried about the impact of plan proposals on the electoral fortunes of their respective parties and the District Planning Committee duly passed this proposal.

Thus, the sustainability criteria that had been used to allocate development in the first instance were completely undermined, an outcome that would not only require a re-structuring of the District's housing policy but would lead to an extra round of consultations in all the villages now identified as suitable for development.[8] Having spent several years honing the allocation strategy, the planners watched it being displaced in favour of a more nakedly political scheme. They therefore mounted a rearguard action to re-instate the sustainability approach at the heart of the plan.

Following the 1997 General Election, some of the political heat was taken out of the issue and the planners were able to re-introduce their original distribution of rural housing allocations. The Deposit Draft of the plan specified that it was appropriate to concentrate development in the key rural centres and remove the 'third tier' of smaller villages from consideration (Aylesbury Vale District Council, 1998). The preservationist approach was therefore re-instated, with the vast bulk of rural villages once again offered protection. The final version of the plan sanctified the process of dispersed concentration in the countryside.

This brief overview of planning in Aylesbury Vale indicates that the distribution of local housing development will be determined by the interaction between preservationist networks seeking to steer development into particular (i.e. urban) locations and the demand projections coming down through a vertically aligned policy network that stipulate how regional housing demand should be met. Thus, local conventions associated with neighbourhood, community and environment come up against demand conventions associated with growth, need and market demand. As the case presented here shows, the interaction between these differing convention types in areas where preservationist networks are strongly represented means that already developed areas (such as Aylesbury) become subject to ever-increasing levels of development, while the small villages, rural areas and green belt are afforded increased protection. In turn, these latter places become the most valued environments in the county and therefore home to groups of people who are ever more determined to retain their protected status. We see, then, within the preserved countryside another form of differentiation taking place: between those places that can be 'let go' to development and those that must be afforded permanent protection. The former are shaped by conventions of demand, the latter by conventions of environmental protection.

The changing context of South East regional planning

The events described in the previous sections show that in their attempts to control the housing development process, preservationist groups are confronted by a development network that attempts to push conventions of demand down from the national to the local level of planning. It is not surprising then that local preservationist actors should come to realise that political decisions taken at higher spatial scales are important in determining the outcomes of their own struggles. In the light of this realisation the local preservationist networks begin to build alliances further up the scales of governance (Murdoch and Marsden, 1995). As a consequence, the politics of preservationism becomes evident at the regional tier of planning. Given that this tier is a key node in the vertical network of housing provision it is worth briefly considering how the arguments tend to unfold at the regional level because they may indicate a possible future for the rural localities of southern England, and thus the 'preserved countryside'.

During the 1990s the regional planning body for the South East was SERPLAN, an umbrella organisation made up of representatives of the region's county councils.[9] Composed of county representatives, SERPLAN had long expressed the anxiety of the counties about ever escalating levels of housing development in the region. This anxiety became particularly pronounced after the government's 1995 national housing projections indicated that 4.4 million houses would be needed nation-wide between 1996 and 2016. Of these, over one million would be located in the South East outside London. Thus, when it came to the review of Regional Planning Guidance in the late 1990s, SERPLAN was looking for a way to register local preservationist concerns by limiting overall levels of development (Murdoch and Abram, 2002).

The review began when SERPLAN, as regional planning body, conducted what it called a 'regional capability study' in order to investigate the consequences of meeting differing levels of housing demand across the region (SERPLAN, 1997). This study was designed, first, to give a comprehensive overview of the region's environment and land resource and, second, to demonstrate the key impacts and consequences associated with accommodating alternative levels and distributions of housing growth (ibid.: 1). When the capability study was published in 1997 it showed that the region could accommodate no more than 914,000 houses if environmental constraints were to remain in place. Thus, SERPLAN felt obliged to challenge the national housing projections which indicated that 1.1 million new households would form in the region between 1996 and 2016. Its counter proposal was that the region should accommodate no more than 800,000 new dwellings during the plan period.[10]

SERPLAN set the regional projections within a vast range of complicating factors, summarised using the language of 'sustainability'. The draft regional strategy was in fact called *A sustainable development strategy for the South East* (SERPLAN, 1998). The main strategic approach in this document was based on a belief that past patterns of development in the region had become 'unsustainable' (e.g. development gives rise to population dispersal, urban sprawl and car-based travel) and it was argued that formulating an adequate response to such patterns was the 'central issue' for the regional strategy. Sustainability criteria (e.g. protection of critical regional assets and efficient management of environmental and natural resources) thus shaped the various policies.

Moreover, at the time that SERPLAN was undertaking this review, Tony Blair's new Labour government was attempting to respond to the debate that had broken out over the figure of 4.4 million new houses generated by the housing projections (DoE, 1996).[11] One key element of its response was an announcement – in *Planning for the communities of the future* (DETR, 1998b) and the revised version of PPG 11 (Regional Planning) (DETR, 2000b) – that regional planning authorities would be permitted more responsibility for deciding the levels and distribution of housing land in their regions. It quickly became evident that SERPLAN was prepared to take up these new opportunities by promoting 'sustainability' criteria in its regional plan. In fact, the draft regional strategy placed the discourse of sustainable development at the *centre* of regional planning policy and reduced the number of houses to be provided over the plan period accordingly.

In challenging the national housing figures from the standpoint of 'sustainable development', SERPLAN was strongly supported by preservationist networks who saw SERPLAN's shift to a sustainable regional planning as broadly reflecting their own environmental concerns. The leading protagonist here was the CPRE, a group that has long been at the forefront of efforts to counter sprawl, ribbon development and suburbanisation (Matless, 1998). The CPRE has been one of the few environmental groups to campaign consistently on housing issues (Lowe *et al.*, 2001) and during the 1990s it focused a great deal of attention on the role of national housing projections in pushing new housing development into countryside locations. It argued that the 'household projections have a profound effect on decisions made about new housing development' and asserted that the figures are 'the fundamental driver of the planning for housing process' (CPRE, 1995: 132). The CPRE campaign ran parallel to the review of RPG in the South East region. The group therefore established an informal alliance with SERPLAN and all the local amenity societies worried about the implications of the housing numbers.[12] SERPLAN's rejection of the national housing projections was fully supported by the CPRE and other such groups.

Development interests, however, took exception to SERPLAN's approach. For instance, the House Builders Federation (HBF) argued that 'ignoring projections of household growth will leave planners with no basis for planning and will ensure that the national priority which housing must be given . . . will be ignored' (HBF, 1998: 3). Such concerns were echoed in comments from the

Town and Country Planning Association (TCPA), which argued that downgrading the projections would have serious consequences for the operation of the housing market in the region. Peter Hall, a leading member of the TCPA and former government advisor on planning policy, spelled out the consequences in stark terms when he said: 'the poorer one-person households – including young people who have left the parental home, divorced and separated people, and older widowed people – may be left effectively homeless' (1998: 200).

These arguments came to a head during a public examination of the South East RPG overseen by a government-appointed Panel (chaired by a retired planning inspector, Stephen Crow). A number of 'stakeholders' – including the CPRE and the TCPA – were invited to discuss the draft RPG and following this discussion the Panel published its own report (in the autumn of 1999). The Panel report effectively provided a critical assessment of SERPLAN's attempt to shift the balance of policy considerations towards sustainable development.

To begin with, the Panel professed itself unconvinced by SERPLAN's rejection of the household projections, largely because it regarded the alternative approach sketched out in the draft RPG as incapable of meeting regional demands for housing and jobs. The Panel argued that 'the essence of planning lies in taking a view of what is likely to happen in the future and planning to meet it' (Report of Panel 1999: 47). It therefore urged SERPLAN to 'adopt a reasonable and responsible approach to future housing requirements, taking due account of the household projections' (ibid.: 48). The Panel rejected SERPLAN's view that the projections are flawed and challenged the assumption that the required level of housing would 'unacceptably compromise other objectives', notably those associated with sustainable development (ibid.: 48). It also expressed the view that no more than 50 per cent of new housing could be built on brownfield land and therefore suggested that this lower figure be adopted over the government's preferred target of 60 per cent.

In effect, the Panel re-asserted conventions of demand over those associated with 'sustainable development' (Murdoch and Abram, 2002). It argued for a continuation of past growth trends, with some limited attempt made to channel economic activity away from the development hotspots in the west of the region (around the so-called 'M4 corridor') towards the more deprived and unattractive areas of the east (the so-called 'Thames Gateway'). This prioritisation of economic demand conventions led preservationist networks to conclude that the Panel report essentially challenged the whole basis of sustainable regional planning. For instance, the CPRE said:

> The RPG should . . . recognise the limited environmental capacity of the region as a whole in terms of demands on natural resources, such as water, the effects of traffic growth and impacts on landscape and wildlife and key assets such as rural tranquility. . . . Effective and regular assessment of the region's environmental capacity and the means to monitor the impact of new building so that capacity constraints are respected are . . . important. Without such management measures, greenfield land will inevitably be developed in

> preference to re-cycled urban sites, driving dispersion rather than responding to need, and fatally undermining the Government's economic, social and environmental aims.
>
> (2000: 4)

Following the Panel Report, the CPRE ran a high profile campaign in the national, regional and local media (targeted on the broadsheet press). It held a number of meetings with civil servants, special advisors and ministers and put pressure on the House of Commons Environment Committee which was holding an enquiry into the whole issue of planning for housing so that the Committee 'came up with the right conclusions' (as one CPRE activist put it in interview). In all these policy arenas the CPRE presented an alternative set of planning policies for the South East based, it was argued, on the 'sustainable' approach officially endorsed by the government in *Planning for the communities of the future* and implemented by SERPLAN in its original draft of the RPG. In essence, the CPRE argued that planning in the South East region must adopt the 'brownfield first/ greenfield last philosophy' so that new housing is concentrated in urban locations.

These arguments appeared to win favour with the Labour government and the Environment Secretary, John Prescott, announced in March 2000 that he was rejecting the Panel's amendments. He re-iterated that he was moving away from 'old style predict and provide' planning for housing (as illustrated, he argued, by the Panel report) and claimed that he was opting instead for a 'more flexible and responsive approach based on planning, monitoring and managing' (DETR, 2000a). Prescott proposed that, rather than providing the full number of new homes specified by the Panel, only 43,000 new dwellings would be needed in the South East each year between 1996 and 2016 (DETR, 2000c). If this figure is aggregated up for the whole period then it means 860,000 new homes in the region (a figure, it should be noted, that lies expediently between the SERPLAN figure and the national forecasts), although it was emphasised that such an aggregation would be to 'misinterpret' the 'plan, monitor and manage' approach (GOSE, 2000: 11). According to this approach, planning authorities in the South East should plan to meet 43,000 additional dwellings per year, subject to regular review. They should aim to monitor the effect of this provision against a series of indicators, manage it and, if necessary, adjust the rate of development in the light of the monitoring. Hence, 'the annual rate of 43,000 net additional dwellings is not fixed for twenty years' (ibid.).

In broad terms, this announcement seemed to concur with the preservationist view. It aimed to place the philosophy of 'urban first/greenfield second' at the heart of the regional plan and sought to reduce the significance of the overall housing figure. However, SERPLAN and the preservationist groups still claimed the amount of development projected by the revised guidance would breach environmental constraints in the region. These arguments again appear to have had some influence on government for in its final RPG document, published in March 2001, the regional housing figure was reduced even further, down to 39,000 per annum for the first five years of the guidance period. Even the overall number of houses to be built after 2006, though given an indicative target of 43,000 per annum, was not to be rigidly adhered to:

> Recent housing completions in [the South East outside London] have averaged about 39,000 dwellings a year . . . the rate of completions should be increased in future years. However, until assessments have been completed of the capacity of urban areas and the scope for the potential growth areas to accommodate additional development, it is premature to specify precisely the increased level of provision and how it might be distributed, although it would be expected to result in around 43,000 dwellings a year. Future development needs to be achieved without perpetuating the trend to more dispersed and land-extensive patterns of development . . . It should be possible, through a plan-led process, to provide more dwellings than have been provided in recent years with proportionately less impact on land and other resources.
>
> (DETR, 2001: 47–8)

Environmental and localistic conventions thus came to lie at the heart of the South East RPG. The final document states unequivocally that: 'Future development is to be accommodated within all parts of the urban areas of the South East, both within central areas and the suburbs, in order to make better use of land' (ibid.: 19). In other words, the regional plan should actively work against the dispersal of new housing. In short, a firm commitment to protect the rural areas of the South East has been set within regional planning policy.

Conclusion

The example of planning-for-housing in the South East shows how the struggle to control territory takes place in the preserved countryside. We see here an illustration of Hoggart's (1997) view that local preservationist networks are frequently confronted by well-resourced, national-to-local networks. These national-to-local networks seek to push demand conventions into local areas in an attempt to dislodge the anti-development coalitions that have built up around the local planning system. Put simply, the local coalitions are more concerned with the protection of green space than the building of homes to meet demand, while the development networks are more concerned with supplying local houses than with the quality of local environments. Each network therefore asserts a differing combination of conventions and these combinations become embroiled in planning debates at county and district levels.

In the two planning case studies presented above we find the two networks struggling over the levels and distributions of housing development. The local network of politicians, residents and amenity groups coalesces around spatial distribution policies and seeks to ensure that strong urban concentration/rural protection approaches are put in place. In part, this approach reflects the environment network's desire to secure the (aesthetic) preservation of 'their' rural spaces.

However, it also stems from a more altruistic politics of local preservationism, one that seeks to ensure that the whole local area is protected from over-development. In the resulting dispute over housing allocations, urban areas are seen as the most appropriate locations for new development, while rural areas are thought worthy of strong protection. Where the distribution accords with this spatial logic, it indicates not only the strengths of the competing networks but also how conventions become spatially dispersed, with conventions of demand bound into the fabric of urban areas while conventions of domestic worth and environmental value are embedded in the protected landscapes and villages of the countryside.

In the local study we found both networks making headway in the local plan reviews: the distribution of housing accorded with preservationist aspirations; the overall numbers of houses to be accommodated conformed to national housing projections. This local struggle over housing development is, however, currently being re-cast by changes in the overall policy framework, notably the growing strength and autonomy of regional policy processes. In the wake of this change, local preservationist networks begin to coalesce with national environmental networks so that a preservationist coalition emerges across the local, regional and national scales of governance. This local-to-national network takes advantage of recent innovations in regional planning policy in order to bring conventions of localistic protection and environmental worth to bear. Thus, preservationism is beginning to assert itself higher up the scale of governance, even extending all the way up to national government and the Secretary of State for the Environment.

The upshot is that a regime shaped by the conventions of 'aesthetic reflexivity' is now beginning to encompass the rural areas of South-East England. This regime, we can fairly confidently predict, will bolster the politics of preservationism at the local level. The consequence is that the 'preserved' countryside may become a stable part of the South-East region. While this might seem a surprising outcome in an area marked by high levels of economic growth, it can also be seen as illustrating the strength of those middle-class preservationist networks that have emerged in the wake of counterurbanisation and which are determined to ensure the countryside's continued protection in line with their own aesthetic aspirations.

5 The 'contested countryside'

Introduction

We now turn to the 'contested countryside' which we previously described as located in areas where growing numbers of middle-class activists confront a well-entrenched set of developmental actors (who are still well represented in local political structures) thereby giving rise to increased conflict around land uses (Marsden *et al.*, 1993). Using Devon in the South West of England as an exemplar of the contested countryside type, we assess how developmental and environmental networks come into conflict around rural economic development issues. In contrast to the 'preserved countryside', where preservationist coalitions are seen as representative of the local social context and where development is something imposed via national-to-local networks, in the 'contested countryside' we find that it is the environmental groups that are seen as 'external' actors while development networks purport to reflect the interests of the locality and its traditional rural residents.

In part, the relationship between development and environment networks sketched out in what follows reflects the distinctive circumstances prevailing in the economic development sector. Unlike in the housing sphere, there is no vertical planning network operating in the field of economic development. Indeed, national planning policy has little to say about the allocation of land for employment or commercial purposes. Traditionally, policy has exhorted local authorities not to obstruct economic development unless absolutely necessary and has asked them to keep a range of sites available to meet different types of demand for land. In a recent study of planning for local economic development, Healey (1999: 31) could discern 'little coherence in the expression of national planning policy in relation to site allocation for economic development purposes'. Local planning authorities have therefore largely enjoyed the freedom to determine their own economic development strategies. This situation differs from that prevailing in planning for housing where we found in Chapter 4 that local preservationist coalitions are attempting to protect the countryside in the face of a strong national-to-local housing network. By focusing on economic development in this chapter we therefore examine a much more localised expression of the conflict between developmental and environmental networks.

In the first part of the chapter we show how local politics in much of rural

Devon still works in line with a localistic and communal set of considerations in which development to meet local needs is routinely seen as an intrinsic part of rural life. However, in the wake of counterurbanisation, this situation is changing and local development interests are now confronted by a strengthening environmental constituency, one that asserts the aesthetic concerns that are so valued by new rural residents. We illustrate the growing significance of the new environmental networks through the examination of a leading environmental group, one that is closely associated with the counterurbanising countryside, the CPRE.

The CPRE is a national charity and 'exists to promote the beauty, tranquillity and diversity of rural England by encouraging the sustainable use of land and other natural resources in town and country' (CPRE, 2001: 15). It is organised at national, regional and local scales and is usually regarded as a group that seeks to protect the countryside from unwelcome development. We look in some detail at the Devon branch of this organisation in order to investigate how it involves itself in the contest between development and environment. As we shall see, because the CPRE is a *national* organisation, it is frequently characterised as an 'external' actor by local development interests. Such a charge potentially carries a great deal of weight in the politics of the contested countryside because the CPRE is also thought to represent the new rural residents, the counterurbanisers. Thus, 'internally' generated developmental conventions apparently confront 'externally' constructed environmental conventions in local political arenas. This confrontation shapes the conduct of politics in the contested countryside.

The 'contested countryside' in the South West of England

The South West region, which extends from the buoyant economic areas around Bristol in the north down to the more depressed area of Cornwall in the south, is home to almost five million people, half of whom reside in rural areas – the highest proportion of any region in England (Countryside Agency, 2001). This population has been growing rapidly over the last twenty years or so, largely due to substantial in-migration. Although some of this in-migration has flowed into the urban areas of the north, most of it has headed for the small towns, villages and countryside across the region. During the 1990s the rural population of the South West grew by 3.6 per cent, to reach 2.3 million.

Significant employment growth – of around 10 per cent between 1991 and 1996 – has been both a cause and consequence of this growing population. However, new employment opportunities have been concentrated in the north and east of the region, where the high-tech and business and financial services industries are located and most of the existing employment is found. Further west and south employment growth has been slower, and local economies are more dependent on the primary sector, tourism and consumer services.

Levels of self-employment at 16 per cent are higher in the South West than the national average (13 per cent) and are particularly high in remote rural areas – especially those which are farming and tourism dependent – reaching 18 per cent in Devon (Devon County Council, 1999). The business structure of the rural areas is small scale: micro businesses – that is, those with a workforce of nine people or less – predominate, accounting for 91 per cent of all businesses in the region. Only 1 per cent of businesses employ more than 50 people. The small business structure means that average earnings in the region are lower than for England as a whole. This is compounded in the rural areas by the reliance on traditional industries. Cornwall has the lowest wage levels of all the English counties.

The rural South West also remains largely agricultural. The farm workforce of just over 80,000 is 3.3 per cent of the region's total working population, about twice the national proportion. Nearly 13 per cent of all business sites are classified as agriculture, hunting, forestry and fishing, again about double the national average (GOSW, 2001). This traditional economic structure dominates the landscapes of the region and also shapes local politics and land development processes in rural areas.

Of all the counties in the South West, Devon is the most agricultural. While in recent decades it has experienced significant economic and social changes, it remains predominantly rural in character. Along with its westerly neighbour, Cornwall, it used to be thought of as a peripheral sub-region – the 'far South West' – that was lagging behind in terms of economic development. However, with the growth of the South East and the outward spread of economic growth along the London–Bristol (M4) corridor, so the more accessible southern and eastern zones of the county have seen their economies drawn more closely into line with that of southern England. Thus Devon, which is a large county geographically, has come to straddle the divide between the prosperous south of England and the poor western periphery, which still includes Cornwall and the remoter parts of North and West Devon.[1] Associated with this shift in the economic geography of Devon has been a significant re-structuring of the social and economic complexion of the county, including a steady decline in agricultural and ancillary employment, considerable expansion of the service sector and a high rate of population growth through in-migration.

However, these changes have operated upon and through a rather traditional social and geographical structure, one that remains largely rural. Devon was once the third most populated county in Britain but nineteenth-century urban-industrialisation largely passed it by. As W. G. Hoskins (1959: 158) explains: 'There was no Industrial Revolution in Devon. . . . There were no coalfields to provide cheap power for factories and mills, and Devon remained mostly a rural county'. Most villages and towns declined in prosperity, size and consequence. During the first half of the twentieth century, while the towns of Plymouth, Exeter and Torquay expanded, population fell across more than two-thirds of the county. Devon continued to experience de-industrialisation as tourism and domestic, administrative and retailing services largely replaced traditional industries such as textiles and quarrying. Increasingly, it was the absence of large-scale industry that became Devon's defining feature. In the words of one historian, 'the material results of the Industrial Revolution – factory towns, slum housing, blast furnaces and coal pits – are almost entirely absent from the county' (Stanes, 1985: 108). In the late 1950s, Hoskins could write:

> The Devon towns are mostly small (all except Plymouth) or keep their small-town feeling. Even in Exeter you can see the green fields of the country at the end of nearly every street-view, and people have the cheerful, rubicund look of country-dwellers and not the miserable, slave-like expression that one sees in London and the big industrial cities.
>
> (1959: 167)

Since then, the county has experienced substantial flows of in-migrants, attracted in the main by its rural characteristics. Even now, though, Devon has the relatively low population density of 158 persons per square kilometre compared to an English average of 378. Some districts are exceptionally low, notably West Devon (40), Torridge (56), Mid Devon (73) and North Devon (79). Migration into the county has been very diffuse, reinforcing the dispersed settlement pattern. The establishment of Torbay and Plymouth as separate unitary authorities in 1998 left Exeter, the seat of county administration, as the sole urban centre within Devon. The majority of people live in the towns and villages scattered across the county.

The rural character of the county is also evident in the continued significance of agriculture: in the 1991 Census, three of Devon's districts – Mid Devon, Torridge and West Devon – had 10 per cent or more of their workforce directly engaged in farming and horticulture. The distinctive pattern of farming contributes strongly to the identity of Devon, with pastoral livestock farming (i.e. dairy, beef and sheep) dominating the landscape. The farms are relatively small, averaging 43 hectares (compared with 61 hectares nationally), and the county has the largest number of any in England. Two-thirds are owner-occupied and they rely wholly or mostly on family labour: three-quarters of the agricultural work force is made up of farmer, spouse and other family members (Devon County Council, 1999).

This small-scale, pastoral farming contributes to the distinctiveness and diversity of the Devon landscape. Much of the land has low fertility and can only support extensive grazing but this in turn may help to maintain important habitats and wildlife. The county includes two National Parks, Areas of Outstanding Natural Beauty and Areas of Great Landscape Value. Most of the coastline is designated Heritage Coast. The Devon Biodiversity Action Plan describes 'a cornucopia of species, niches and landforms' and 'an almost unrivalled range of ecosystems' (Devon County Council, 1998a: 1 and 7). The pastoral character of the county, along with its coastal attractions and seaside, lies at the heart of Devon's touristic appeal.

Devon tourism, which has evolved over many years, depends upon the social and environmental image of the county. Through the nineteenth century a number of watering places along the north and south coasts developed reputations as specialised health resorts and attracted a national clientele. Distance from the major conurbations enabled the Devon seaside resorts to preserve a high social tone and maintain a sedate, genteel ambience (Travis, 1993). In the twentieth century, pleasure supplanted health as the dominant reason for holidaying in Devon. Visitor numbers greatly expanded, the resorts became less exclusive, and motoring opened up the whole of the coast and the countryside to tourists. Even so, Devon remained set apart from other holiday destinations in England in relying to an unusual extent on its natural assets to attract its clientele. It was its exceptionally mild sea air, splendid scenery, rich flora and fauna and pastoral charm which made it an attractive destination (ibid.).[2]

The popularity of the county as a holiday destination is matched by its attractiveness to people moving from elsewhere in the UK. Other rural areas and regions have experienced counterurbanisation, but Devon stands out in terms of the volume of the influx and the preponderance of inter-regional (rather than 'within region') migration. Population growth rates have been among the highest in the UK, and are entirely due to net in-migration. The county's population grew by 16.7 per cent between 1971 and 1996, with most of the growth occurring in the rural districts. Growth exceeded 25 per cent in the districts of East Devon, South Hams and Teignbridge. The long-distance migrants and retirees come predominantly from much more heavily urbanised regions, mainly South-East England, London and the West Midlands.

A recent study of counterurbanisation in North Devon by Bolton and Chalkley (1990) showed the new arrivals to be more middle class than local residents. Some of them had come to Devon to retire but for most the move had been associated with work availability. However, when asked about their main reasons for moving into the area, most counterurbanising households cited non-economic factors as the main rationale. These included such things as movement away from pressurised urban life and escaping the 'rat race'. Bolton and Chalkley (ibid.: 39) conclude that incomers to Devon 'tended to leave their former areas for lifestyle and socio-environmental reasons'.

Halliday and Coombes (1995) draw similar conclusions from their 1987–8 survey of counterurbanisation into the county. Although they found employment motivations to be the most common impulse behind households' decisions to relocate, the most widely cited reason for moving to Devon was the 'scenery' (ibid.: 442). Halliday and Coombes conclude that, while employment considerations are important in accounting for a certain proportion of moves, 'motives such as "way of life" and "scenery" provide something more akin to a "frame of mind" within which more specific and rational decision processes take place' (ibid.: 445). In short, people move to Devon in order to live in a 'green and pleasant' environment and to enjoy a more rural pace of life.[3]

During the development boom of the 1980s, the migration flows that had previously impacted on the coastal and market towns began to affect the smaller villages and hamlets of the deeper countryside. Transport improvements, including the completion of the M5 motorway and the north Devon link-road, opened up many parts of rural Devon to longer-distance commuting (to Bristol and the M4 corridor, for example). Residential dispersal was also encouraged by the liberalisation of the planning system and facilitated by farmers, many of whom, with government encouragement, sought to realise some of their assets by releasing land for development or converting redundant farm buildings into expensive dwellings (DoE/Welsh Office, 1992; Kneale *et al.*,1992). The social fabric of once agricultural villages and hamlets was transformed.

The scale of population growth, economic re-structuring and the associated development pressures have all led to environmental issues becoming prominent in local politics. The combination of an attractive landscape, important environmental characteristics, and in-migration have stimulated the formation of local amenity groups and the region has been an important location for the growth of environmentalism nationally within the UK. The Green Party recorded its greatest electoral achievements in the South West, and at times in the late 1980s it seemed on the verge of a major breakthrough in the region, most notably in the 1989 European elections when it received over 20 per cent of the popular vote. National campaigns to halt the ploughing up of moorland in National Parks, to stop the drainage of lowland bogs, to combat river pollution from farm waste and to clean up sewage contamination of bathing beaches first took off in this region, led by a local network of activists and sympathetic journalists (Lowe *et al.*, 1997). Groups such as the Woodland Trust and Sustrans also began in the area.

By and large, the carriers of environmental conventions have been 'incomers'. However, in bringing the environmental repertoire into the region's politics, they have often found themselves challenging established local interests – whether hoteliers seeking to boost visitor numbers, local developers attempting to squeeze development proposals past planning controls, commercial interests and trades people looking to expand particular settlements or farmers intent on intensifying their production and diversifying their holdings. At the core of such disputes are basic disagreements about the well-being of the countryside and its inhabitants. For example, the influx of large numbers of newcomers in the 1980s helped catalyse a major shift in public attitudes to agriculture and the countryside. Many farmers now had new neighbours – often retired migrants or well-to-do professional or business people – with quite different perceptions of the function of rural areas. As a result, one in six Devon dairy farmers surveyed in the early 1990s had experienced direct pressure from neighbours and local people to change their farming practices (Lowe *et al.*, 1997: 155).

Yet, in opposing such a traditional local network, the environmentally conscious incomers have also re-confirmed the pre-eminence of the 'rural' character of Devon for the pastoral countryside is now seen as central to any form of rural preservationism. The Devon Biodiversity Action Plan, for example, while recognising that some modern farming practices are not especially sympathetic to local environments, nevertheless concludes that, 'farmers and land managers are central to the goal of maintaining a rich and varied natural environment in Devon. As stewards of most of the land surface in the County, it is they who ultimately control the future' (Devon County Council, 1998a: 45). Thus, while challenging traditional local development and farming interests, environmental groups have also had to reach an accommodation with them in order to ensure that the distinctive rural and environmental character of the area is maintained.

Environmentalists and rural development interests do therefore share some common ground, with one network recognising that the maintenance of the countryside relies on a viable and sympathetic agricultural industry and the other accepting that farmers have a social responsibility as guardians of the countryside.

They concur also in valuing the countryside, not just as a physical and natural entity, but as the setting for a traditional, non-urban way of life. Yet, despite these commonalities, there remains much mutual suspicion. Environmentalists suspect that, left to their own devices, diversifying or intensifying farmers would damage the environment. For their part, rural development and farming interests suspect that, given their head, environmentalists would fatally constrain the opportunities of farmers, landowners and developers to pursue new commercial possibilities.

The divisions that lie between the environmental and developmental networks thus rest upon different appreciations of the core values of the farmed countryside and the major threats it faces. On the one hand, the environmentalists are concerned about the damage of inappropriate development in the countryside; they thus assert ecological conventions in order to bring the rural areas of Devon within discourses of environmental valuation. On the other hand, farming and rural development interests are concerned about the options and choices available to rural families as they seek to sustain their livelihoods; they therefore assert conventions associated with growth, market demand and rural development. At its root, then, the disagreement between the two networks is a struggle over the conventions that are to govern development processes in the countryside.

Agrarian politics in rural Devon

Traditionally, farmers and landowners have been the dominant social group in Devon politics (Stanyer, 1978; 1989). In the 1960s almost 40 per cent of the County Council belonged to the landed and agricultural interests. Farmers were a substantial minority on many parish councils and were dominant in the rural district councils. Although not such a dominant force today, farmers are still disproportionately represented at the various levels of local government. Stanyer suggests that the farming community has traditionally played a systemic role in Devon politics, in furnishing the sort of leaders who could and would fill the roles created by the local political system:

> In rural areas farmers are particularly well-suited to grow into such roles; they have the time and money to serve on councils and committees, including those at two or more levels, some will have a local social status, of which 'outsiders' may be unaware, which makes them obvious local leaders, and many will have been involved in the activities of their branch of the National Farmers' Union: a sophisticated and government-oriented body.
>
> (1978: 282)

Processes of political recruitment, socialisation and promotion tended to reinforce the representation of farmers in the political system. Electoral divisions favoured the rural areas, where farmers were established social leaders, and electoral behaviour showed greater stability in such areas, which gave farmer-councillors added advantages within the system. First, serving members were more likely to seek re-election in rural than in urban areas. Second, elections were less likely to be contested. And, third incumbents were less likely to be defeated if opposed. These processes not only disproportionately retained farmer-councillors within local politics but also gave them experience and seniority, which heightened their chances of being recruited to other public bodies (including higher tier councils) and to council leadership positions. A further crucial factor was the absence of partisanship: in rural areas at both district and county levels there was almost militant opposition to party politics in local government (Grant, 1973). Thus, the local connectedness of candidates carried considerable weight.

The rural character of Devon has therefore traditionally facilitated the consolidation of a political network oriented to farming and rural development concerns. In consolidating itself, this network has emphasised the communality of interests and identity that arise through local attachment. A potent form of such localism is *agrarianism* – an ideology that places farmers at the heart of the rural community and identifies the well-being of the countryside with the well-being of agriculture. Central to agrarian thinking is the perpetuation of a family form of farming maintained through strong kinship networks and an attachment to a country 'way of life' (Williams, 1963).

At the heart of many of Devon's rural communities are tight networks of longstanding farming families. They are bound together by reciprocal relations of varying kinds, underpinned by what D. M. Winter (1986: 245–6) calls 'a moral economy in which notions of "good farming", kinship, localism, and religious sensibilities all play their part'. Winter argues that this ideological unity has been functional to the survival of family farming by enabling extensive, informal exchange and cooperation between farming neighbours and relatives that allows farms to cope with seasonal, economic and family cycles. More broadly, in identifying the well-being of the countryside and the rural community with agriculture, agrarianism underpins the social and political leadership of farmers in political networks.

Yet, as the proportion of farmers has diminished, and as the county has been socially transformed by in-migration, so the farming community has gradually become less dominant in Devon politics. A crucial development in this regard was the re-organisation of local government in the 1970s. This brought the unitary authorities (county boroughs) of Torquay, Plymouth and Exeter into the county system as districts (a process reversed for Plymouth and Torquay in 1998) and it amalgamated all the small town and rural districts into seven new district authorities. One effect was to carry the party politics of the urban areas out into the rest of the county, a move that effectively ended the involvement on the County Council of the traditional type of social leader – such as prominent members of the gentry and the landed class. Since then, Devon's County Council chairmen have included estate agents and hoteliers, that is, people chosen by their party groups.

The new district councils drew in the local farmers who had dominated the old rural districts and the business and trades people who ran the coastal and market towns. And at the district level in the rural areas there was resistance to the drive for partisanship. This was particularly marked in north and west Devon – there independents fought off the (mainly) Conservative challenge (Stanyer 1989: 37) – and independent councillors (many of them farmers) remain active in rural Devon politics. In North Devon, for example, throughout the 1990s, roughly one in five councillors were farmers, mainly independents (Gilg and Kelly, 1996). Some of the same factors are at work as before: electoral divisions still favour the rural area; electoral behaviour shows greater stability in the traditional rural areas; and the rural parish structure which often serves as a recruiting ground for potential councillors has remained largely untouched through successive bouts of local government re-organisation. All of these factors still serve to channel local farming leaders into service on the district councils in the more rural parts of the county.

Agrarianism in local politics seems to have fused with a more general localism which, along with business autonomy and family values, is a key part of the value orientation of the petite bourgeoisie who make up the local economy of much of Devon. Thus, across many parts of rural Devon, a development-oriented localistic network remains in control of political institutions and tends to push conventions which link together local well-being and economic development. This is illustrated by the fact that, at the time of writing, there were 106 independent councillors compared to 110 Conservatives, 106 Liberal Democrat and 29 Labour in the district councils. This political complexion reflects the rural and small town character of the county, the predominance of small businesses and the continued significance of agriculture as a source of social and political leadership. Many of the independent councillors, in particular, identify strongly with local farming, business and community interests. They thus articulate rural development repertoires refracted through a discourse of localism: that is, local jobs and housing should be provided for local people in a context where wage rates are low, job opportunities restricted, and where incomers are buying the available houses and pushing up prices.

However, as we shall show in the next section, in articulating this repertoire, local councillors are increasingly confronted by the environmental network that has emerged on the back of counterurbanisation. Thus, on the one hand, there is a recognised need for rural development and farm diversification to boost local incomes, support rural services, and offset the impact of the farming crisis; on the other hand, there is great pressure to protect and preserve Devon's environmental heritage. This leads to intense debate and occasional conflicts about how to manage and regulate change in the county in a way that will satisfy the two competing networks and their respective constituencies.

Environmental politics in rural Devon

We can illustrate the nature of the growing environmental challenge in rural Devon by examining the make-up of an environmental organisation that is active across the county, that is, the CPRE. This organisation is significant for two reasons. First, it has long campaigned to protect the countryside against unwarranted development and is organised on a county basis to allow protectionism to be pursued in a structure that matches local government and the planning system. Second, the growth in membership of the CPRE is closely linked to counterurbanisation and the CPRE is generally strong at the local level where counterurbanisation is strong (Lowe *et al.*, 2001). By focusing on the CPRE in Devon we can thus gain some considerable insight into the relationship between in-migration and rural environmentalism. In the next section we turn to look at the contestation between the CPRE and local political interests in the context of local policy processes in order to show how networks of developmentalism and environmentalism confront one another in Devon.

The Devon branch of the CPRE was established in 1955, during a period when planning and development-control functions were under the control of the County Council. At that time it operated as a county-level committee bringing together representatives from various organisations in order to press for better planning. In short, it was a force for the *professionalisation* of planning in the county.[4] In its early years it gathered together a small group of supporters, mainly landowners and professional people. By 1971 the Devon CPRE had a membership of 200. Support for the branch grew rapidly in the 1980s when many of the villages and small towns faced unprecedented levels of housing development. A number of new residents became involved at this time, often because of a single development proposal. As one member put it, 'local people often don't have the facility to put forward a good case'. As a result, incomers frequently found themselves spearheading anti-development campaigns.

The influx of new residents prepared to take a lead in local amenity politics led to the establishment of district groups (which now cover the whole county – North Devon, South Hams and Plymouth, West Devon, East Devon, Torridge, Teignbridge, Torbay and Mid Devon). One member described the formation of one district group in the following way:

> My involvement with CPRE started in the 1970s if I remember right, when there was a planning application for a field opposite where we lived and there was a lot of local protest and one of the residents in the village suggested that I ought to get in touch with CPRE and see if they would help us. I did this. They did help us. And I thought it incumbent on me to join as a result. We then supported CPRE by attending the various functions which they ran and as a normal subscription paying member. Later on, and I can't remember the exact date, ten or eleven years ago, there was a planning application for a meat packing factory in a village a few miles away and we were so incensed by this that we wrote to a local paper and a resident in the village in question then got in touch with us and said that they were thinking of setting up an East Devon group of CPRE and would we like to join in. And so we did.

We can see from this quotation that localistic concerns stimulate an interest in rural protectionism and then lead on to involvement in an environmental network. This seems to be the traditional route to CPRE activism.[5]

The establishment of the district groups focused a great deal of CPRE attention on the activities of the district councils, in particular the councillors. The re-organisation of local government in the 1970s had seen the eclipse of non-partisan leadership of the county by prominent county figures to which the CPRE had previously related. The new district councils were organised, at least to some extent, upon party lines and recruited the local farmers and business people who had dominated the old rural district councils. Facing them, the CPRE's new district groups drew in activists from the various local amenity societies, village action groups and residents associations that had sprung up to protest against the rash of local housing and industrial developments across the county. These concerned actors looked to the CPRE to help bring some order and restraint to what was seen as a development 'free-for-all' and to counter what they felt was a lax and compromised approach to development control by the district councils. As a consequence, the new district structure became an effective part of the organisation.

This effectiveness was heightened in the 1990s when the district groups were given the responsibility of tracking the local development plan review processes that followed the 1991 Planning and Compensation Act. As we noted in Chapter 2, the 1991 Act stipulated that district planning authorities should draw up district-wide plans. This was the first time such plans had been made part of the statutory system. The Act also stipulated that all development control decisions should be made in accordance with the plan (unless 'material considerations' indicated otherwise). This clause seemed to indicate that the district plan would have a new importance in determining patterns of local development. As a consequence, the Devon CPRE began to focus attention on this more local level of planning. As district planning grew in importance so did the district groups within the CPRE. And as counterurbanisation gathered pace, so the district groups gained in strength making the CPRE a much more formidable force in Devon's politics.

Development versus environment in local policy-making

The CPRE's influence is most evident in the planning arena and it is here that it frequently finds itself confronting local development interests. The CPRE nationally has been a persistent advocate of a plan-led approach to development decision-making and the local branch has sought to encourage adherence to national, regional and local policy, along with a more generally professional attitude to planning issues among planning officials and local councillors. As we shall see, there are varying interpretations of 'professionalism' and the 'plan-led approach'. On the one hand, the CPRE sees these aspects as a way of bringing local decisions into line with national policy; on the other hand, local politicians and developers see them as unwarranted interference in their ability to match local developments to local needs and aspirations.

The CPRE is systematically involved in development control monitoring in the county. The district structure allows it to scrutinise all planning applications and to assess whether they contravene plan policies. The CPRE brings any infringements to the attention of local planners. The attitude towards the planners within the CPRE is that, while the officers are often lacking in initiative, they will generally follow a professional approach if the CPRE remains vigilant. As one CPRE activist commented, 'often enough it is a question of motivating the planners to do what they know they ought to do and get working a little bit harder at it than they would otherwise do'. The CPRE members claim that they actively seek to work closely with planners in the context of local policy processes. As one put it: 'We do try to be as constructive as possible, taking into account the parameters in which they [the planners] have to work'.

The main conflict occurs between the CPRE and the political members. As we outlined above, the councillors are often drawn from the agricultural, tourism and development sectors. It appears that, even where they are in a minority, these councillors can exercise considerable influence in planning decision-making. For instance, Gilg and Kelly (1996: 163) in a study of farmer-councillors in North Devon, claim that 'decision-making was in practice biased in favour of the agricultural sector, largely because of the actions of a few key councillors who employed a variety of tactics to achieve their aims'. These authors argue that, with respect to farm-based developments, farmer-councillors are able to use their superior knowledge of the locality, the farming community, farm business requirements and the personal circumstances of applicants to argue persuasively in favour of agricultural interests. Thus, 'a parochial and professional alliance manifested itself, members with similar interests as applicants overtly demonstrating support for the development against the officer recommendation in a substantial proportion of cases' (ibid.: 161).

It is not surprising, then, that in their efforts to get the district councils to prepare and adopt local plans and then to use these plans to guide decisions on individual applications (in line with the plan-led system), the CPRE district groups see themselves as continually having to combat the inclinations of some local councillors. There is a general feeling that many councillors have little understanding or knowledge of planning and that their response to planning issues is determined too much by localistic calculations, parochial instincts or improper personal considerations. This excerpt from a focus group conducted with Devon CPRE members (during the year 2000) gives a flavour of local CPRE views on this issue:

Mark: What I fear is the lack of technical ability, if you like, of planning committees, or the lack of understanding of planning committees. Well, I've actually seen somebody in a planning committee reading a plan held upside down and making comments on it. And I've also seen in planning committees schemes passed without anybody looking at the plans at all and not understanding the scheme at all. Pretty horrifying really.

Paul: And a failure of correlation between the application and the local plan.

Peter: Not understanding the policies in the local plan.

Raymond: It is only in the last two or three years in fact when you have seen the planning applications go through and are listed for discussion at planning committee meetings any reference to the local plan policies.

Andrew: An indication of this problem really quite recently was when the County Council decided to give short training courses to new members of its Planning and Development Committee. And I for one thought 'jolly good they are going to learn more about planning and the environment, architecture, anything that really matters in planning'. But no, these training courses were merely on the legalities of planning and how not to get themselves into a fix legally, but had nothing whatever to do with the environment or amenities or standards of design or whatever. And this I think is the real problem.

Such complaints have led onto official action, and the Devon CPRE has in the past reported local councillors to the Ombudsman for improprieties in making planning decisions. One local planner commented that the CPRE's past 'whistle-blowing' had certainly been 'salutary'. There was still a sense amongst councillors that CPRE is 'watching' them and this reinforced a feeling amongst the planners that things should be done properly. There was no one else locally who would 'blow the whistle' on the councillors, one planner observed.

While the CPRE activists believe local councillors are tied into local development networks, the councillors believe the CPRE activists represent only the new counterurbanising middle class. Councillors refer to the CPRE branch as 'narrow and anti-development', as the mouthpiece of middle-class incomers or of retired people who don't want to see change. A farmer-councillor referred dismissively to the 'Council for the *Ossification* of Rural England', adding pointedly that he himself was born and bred locally. Another admitted that his colleagues tended to dismiss the CPRE because 'they feel that the CPRE is not a broad church of people . . . you know they are representing the views of a certain section of the community . . . well, they are the middle classes trying to protect themselves'.

In general, there is a division running through the political arena in Devon: on the one side are local councillors – who are characterised as either 'parochial' (by, for instance, CPRE members) or as 'sensitive to local needs' (by other local interests); on the other side are CPRE activists – who are characterised as 'interfering NIMBYs' (by, for instance, some local councillors) or as 'professional' and 'strategic' participants (by, for instance, some planners).[6] However, the nature of this divide between ('local') developmentalists and ('incoming') environmentalists, and the respective strengths of the two opposing networks, varies according to the local context. Thus, in some districts the environmental associations are thought to reflect a legitimate segment of local opinion; in other areas, they are still seen as out of kilter with prevailing aspirations for development to meet local needs for employment and housing. The local context, which is in turn linked to the social composition of particular rural areas, notably levels of counterurbanisation and the strength of the traditional rural population, determines, to a significant degree, the success of environmental actors.

The districts of Torridge and East Devon offer contrasting cases. In the former,

Torridge, there is effectively a stand off between the district council and the CPRE. A senior planning officer characterised the local CPRE as standing 'for a ban on anything', adding 'I don't have any time for them because they go too far'. He did not think their views had support among the councillors (most of whom are independents): 'I don't detect any strong sympathy with the CPRE'. The small Torridge District Group of the CPRE (it has seventy members) finds itself, according to one activist, fighting against an official outlook which believes that the 'enhancement of prosperity . . . begins with allocation of residential and employment land'. In East Devon, in contrast, a senior planning officer described the CPRE as the 'best organised and most well-informed group we have to deal with'. He added, 'they can be allies, depending on the issues', and they were also helpful in 'promoting public understanding of planning'. The CPRE has strong links with several councillors and, in general, council members respect its viewpoint. Yet even here any notion that the CPRE *stands for* rural East Devon is rejected. In the words of a senior planning officer: 'Councillors don't want it to have special status over and above other groups. They often object if we give any prominence to CPRE's views. They say it is just a pressure group'.

Beyond contestation? Linking agrarianism and environmentalism

In Devon counterurbanisers move into the area and tend to hold views that are conditioned by the new uses that they are making of this rural space. These uses may be economic, but are much more likely to be social and environmental. Thus, the new residents will view the countryside through an aesthetic lense and will value it as a social and environmental good, one that should be maintained in an unspoilt fashion for future generations. Conventions associated with community and environment therefore accompany counterurbanisation and become consolidated in the political networks that counterurbanisers become involved in. In Devon these conventions have been pushed pre-eminently by the CPRE. Not surprisingly, given the pattern of population growth in the county, the CPRE in Devon has grown quickly over recent years. This growth has allowed the local organisation to become both more comprehensive and more effective, especially at the district level where many of the most important planning and development control decisions are now taken.

As we mentioned above, the CPRE in Devon has become a force for 'professionalisation' in local political decision making. However, despite this broad aim, the Devon CPRE still finds it difficult to establish good working relationships with councillors across the whole county. This difficulty is most pronounced at the district level where working practices and political cultures vary quite widely. Even in those districts where the CPRE is strong we heard many complaints from CPRE activists about 'time-serving planners' and 'parochial councillors' and from local councillors about 'incomer interference'. In general, the localistic orientation of Devon politics makes it difficult for CPRE members to bring their professionalised and strategic perspectives to bear. In this context the CPRE is seen as an 'external' group, one that has no understanding of the locality and the needs and aspirations of its traditional residents. The CPRE appears to bring with it a set of national or professional environmental conventions and these act to marginalise the more localistic aspirations of Devon's long-standing residents.

Underlying the division between CPRE activists and local councillors are contrasting perspectives on the countryside. For the latter – particularly those who are locally born and bred – the countryside will tend to be a familiar, taken-for-granted reality, one that more easily accommodates most aspects of their lives (living *and* working) in ways that do not threaten its fundamental integrity (e.g. they do not perceive sharp environmental limits that must be observed). Thus, rural development is seen as a process that fits in with the overall ethos of country life. CPRE activists tend to believe that the countryside should somehow be 'set apart' from the processes of development and modernisation that configure other (urban) places. They see in Devon a rich natural heritage that requires protection.

Yet, these characterisations do not fully capture the dynamic nature of the interactions between the networks. For instance, one CPRE activist in Devon, reflecting on his first involvement in amenity politics, said that the campaign to oppose a large commercial development on the edge of the village in which he had recently retired 'had a great effect in uniting the village bringing together newcomers and established residents'. He claimed a local farmer who farmed within the village had thanked him afterwards for all his efforts. Moreover, local CPRE activists also display a great deal of concern for 'local' issues to do with farming and the rural community, as though they too want to reach over the divide between 'incomers' and 'locals'. While their main area of activity remains planning, they are trying to broaden their political range, notably into social housing, agriculture and rural development, to demonstrate a commitment to the well-being of the countryside as a whole. In other words, there is an effort to embed the more 'reflexive' and 'aesthetic' concerns of the CPRE in the agrarian circumstances that prevail in rural Devon.

Thus, both agrarian and environmental networks 'mix up' conventions associated with development, environment and locality. The main distinction between the two is the way they place these conventions into differing hierarchies: on the one hand, the environmental network subsumes economy and locality under a general concern for the environment; on the other hand, the developmental network subsumes the environment into a localistic discourse of rural development. It is the weighting given to local and environmental conventions and the contrasting fashion in which these modes of evaluation are utilised that underpin processes of contestation in rural Devon.

Conclusion

In this chapter we have proposed that any understanding of the contested countryside must link political processes to the impact of population change on rural areas. We have established this link through an analysis of the CPRE in Devon, an organisation that has grown in line with counterurbanisation. Thus, the CPRE is strong in those parts of the county where counterurbanisation is most advanced and there is recognition that many local residents share its concerns. However, in other areas the organisation is weak and it runs up against rural interests who retain a significant political presence and social base and articulate a developmental outlook that derives from local traditions.

The key (though hardly unexpected) finding is that the environmental network represents the most 'middle-class' areas of rural Devon, while the development network represents areas still dominated by farming and small business interests. This not to say that these networks act solely in defence of their respective constituencies; rather, it is to argue that they articulate their concerns and aspirations in ways that largely correspond to those held by given social formations. As we have indicated above, these concerns and aspirations can be characterised as, on the one hand, broadly 'aesthetic' and 'environmental' – that is, they arise from a perception that the countryside is a space to be set apart from the crude economic rationale that determines the development of other (notably, urban) spaces – and, on the other, broadly 'economic' and 'developmental' – that is, development is necessary to meet the material needs of the working rural population.

There is seemingly an endless process of contestation in train between these two perceptions and their associated conventions. Yet, as we noted in the last section, this is not the whole story. Frequently the environmental network incorporates long-standing local residents in opposing new developments, while development interests often assert strongly localistic or environmental concerns. It is therefore necessary to qualify the view that only counterurbanisers value the environment. Moreover, the CPRE is seeking to build up its network linkages with farming and other traditional rural organisations. In fact, the Devon CPRE, along with the counterurbanisers it represents, is looking for ways to assert a more distinctive identity, one that reflects the broad set of concerns that characterise life in rural Devon. In this respect then, we begin to see the contours of a politics which moves beyond contestation and which seeks to bring together rural development and a protected rural environment (rather than pitching these two views against each other). Given that both these perspectives are being proposed by local networks of actors, there is seemingly scope for some measure of network interaction at the local level. Some form of coalition building, although being tentatively considered at the present time, holds the potential to combine the aesthetic and environmental concerns held by the new rural residents with the economic and social aspirations of traditional rural residents. If this potential is realised the 'contested countryside' will remain a distinctive rural space, although its distinctive character will no longer be defined by 'contestation' but by a fusion of agrarianism and environmentalism.

6 The 'paternalistic countryside'

Introduction

Constructing the countryside discussed how the local distribution of property rights, and particularly the private ownership of land, remain crucial factors influencing the patterns and processes of rural development. The general national trend throughout the twentieth century was a shift from a landlord–tenant system of agricultural land tenure to one of owner-occupation, partly fuelled by reforms of tenure and taxation law which systematically worked against the interests of landed capital. Even so, there has been a considerable lag in changes to property relations. *Constructing the countryside* suggested that:

> One of the key characteristics of landowners throughout British history has been their ability to defend and then to adapt their interests in response to changing economic and social circumstances. This means that, at the local level at least, they have often been able to maintain a not inconsiderable presence, even if changes in the manner in which property rights are held and their extent have been profound. There remains a surprising degree of continuity in family ownership, the relatively closed landowning oligarchy of the 1890s having been replaced by a larger but still powerful group, consisting mainly of owner-occupying farmers (including many former landlords) in the 1990s.
>
> (Marsden *et al.*, 1993: 70)

Thus, although the 'single most important feature' (ibid.: 71) of rural landownership in the past 150 years has been the continuous decline in the economic power of the rural landlords, within this overall trend there has been marked geographical variability, which this chapter highlights.

Constructing the countryside proposed that the paternalistic countryside can be identified in those rural localities where large private estates and big farms dominate the land-tenure pattern and the rural land-development process is still shaped significantly by the interests and outlook of existing landowners. Recent years have seen efforts by large landowners to diversify their economic activities or to release some land for development, although their consolidated ownership of large tracts of land makes them less dependent on external developers than farmers with more modest holdings. In the 'paternalistic countryside', estate management and investment decisions may reflect a compromise between financial and social conventions with the deliberate retention, for example, of tied housing and local employment sources.

The continuity of paternalistic relations in rural areas is illustrated in this chapter by the case of Northumberland, a county in the far north of England where landed estates still predominate. We assess the role of property rights and the large landowners holding those rights in maintaining a stable and rather traditional socio-economic formation in the county. We outline the main conventions that are bound into this formation and indicate how a new form of paternalism may be emerging. In so doing, we also consider how counter-urbanisation and environmental networks are excluded as landowners help to maintain not only a traditional *landowning* structure but also a traditional *social* structure. In particular, we consider how the restraints on development exercised by paternalistic landowners prevent forms of middle-class reflexivity being expressed in this rural region.

In short, we ask three main questions. The first is how can we account for the persistence of a significant landed society in rural Northumberland? Second, how has the structure of landownership influenced the evolution of the Northumberland countryside? Third, how does Northumberland's social and political structure influence the functioning of the planning system and the activities of environmental networks? By attempting to provide answers to these questions we are able to assess the significance of paternalism in the contemporary countryside.

The 'paternalistic countryside' in its regional context

The popular image of the English countryside is one of thatched cottages, picturesque villages, parkland, small fields and rolling downs. In contrast, the popular image of the North East of England is of its industrial cities and its rather wild and untamed rural landscapes. In many respects, then, the countryside in this region, with its disused industrial areas and extensive natural habitats, sits uneasily within the English rural context, raising quite distinct issues that are related not just to a more rugged landscape and a harsh climate but also to very particular social and economic concerns.

Bounded by the Pennines, the Scottish Border, the North Sea and the dales and moors of North Yorkshire, the North East region covers an area that includes the counties of Northumberland and Durham and the Tyne and Wear and Tees Valley conurbations. The countryside is very varied, extending from upland moors and forests to agricultural lowlands and coastal dunes. Much of it has a stunning but often bleak grandeur. There is a rich diversity of wildlife habitats and places with strong historical associations. The need to conserve this environment is recognised in a National Park, two extensive AONBs, a Heritage Coast as well as numerous Sites of Special Scientific Interest, scheduled Ancient Monuments, Listed Buildings and Conservation Areas. Some of these features are of international significance: Hadrian's Wall is a World Heritage Monument; the

Lindisfarne area is recognised as a wetland of international importance under the Ramsar Convention; the whole coast qualifies as a Special Protection Area under the European Birds Directive; and the North Pennines is an International Biosphere Reserve. There are also degraded landscapes, associated mainly with industrial and mining decline, where environmental renewal is required to improve the quality of life and to attract new investment. In some places, such as at Druridge Bay, dramatic new semi-natural environments have been engineered out of old mining spoils.

The demise of its former industrial base is reflected in the region's economy. Average earnings in the North East are the lowest in England (Countryside Agency, 2001) and GDP per head in 1998 was less than £10,000 compared with a UK average of £12,500. The industrial legacy can also be seen in an urban concentration of the population: of the 2.58 million total inhabitants only 255,000 (10 per cent) reside in rural areas, far lower than the English average of 28.5 per cent. Although the rural population is growing, this growth is far less pronounced than in other English rural locations: between 1990 and 1999 the population of the rural North East grew by only 0.9 per cent, compared with the English average of 7.58 per cent. Thus, compared with elsewhere in England, the population pressures on the countryside are not so acute.

Small businesses predominate in the rural economy of the region: 92 per cent of all business sites employ fewer than 10 people; and only 0.8 per cent of rural business sites employ more than 50 people (Countryside Agency, 2001). The rural economy retains a traditional, 'old economy' structure. Wholesale and retail, transport, manufacturing and construction are the main sectors, and agriculture, quarrying and forestry remain significant sources of employment, particularly in the remoter parts (Whitby *et al.*, 1999). The information technology infrastructure of the rural areas is relatively undeveloped and in most parts of the region there is a weak presence of the professional and business services firms that tend to dominate buoyant economies elsewhere in rural England (Northern Infomatics, 1997; Raley and Moxey, 2000).

The quality of employment and the income levels reflect the traditional structure of the rural economy. Average earnings in the rural districts stood at £270 per week in 1998 compared to an English rural average of £330 per week (Countryside Agency, 2001). In the light of these structural features, there is an understandable focus among politicians and government personnel on the economic and developmental requirements of the region (see Lowe *et al.*, 2001). Because the amount of counterurbanisation pressure is much lower than in our other two study areas, *urban* development is not commonly perceived to represent a major threat to the countryside. On the contrary, the relatively depressed state of the local and regional economy means that the county actively seeks to attract firms, skilled labour and visitors to rural as much as urban locations.

The social structure of the region underpins this developmental perspective. The economy of industrial Tyneside has produced a large working class and a small middle class.[1] Moreover, compared to the other metropolitan areas of England, the area has not been a significant 'net exporter' of middle-class migrants into rural areas.[2] The commuter hinterland of the Tyneside conurbation is small and constrained – largely to the west (up the Tyne Valley) towards the market town of Hexham (where the transpennine A69 ends as a dual carriageway), and to the north (to the county town of Morpeth where the A1 is likewise reduced to a single carriageway). Beyond this more accessible zone – which covers no more than a fifth of the county and which includes ex-industrial and mining communities as well as more typically suburban countryside areas – Northumberland remains a county where the middle class has established only a limited presence. In part, as we shall discuss below, the lack of a strong middle-class presence can be ascribed to the influence of the large estates (both public and private) that dominate the regional countryside.

The lack of urban pressures on the countryside also means that the politics of the area retains a traditional character. In general terms, the Labour Party holds the key strongholds in the former mining and industrial areas of south east Northumberland (where the trade unions are still a major force in local politics), while the Conservative Party and Liberal Democrats have bases in the rural part of the region.[3] Given the traditional make-up of rural Northumberland, it is perhaps not surprising that environmental networks have struggled to establish themselves. The CPRE, for instance, has found it difficult to organise in the region. As we shall indicate below, the absence of environmental networks and their associated conventions can be ascribed in part to the continued practice of paternalism by private and institutional landowners alike. We now turn to examine in some detail the activities of this distinct social group.

Landed estates in Northumberland

For centuries, Northumberland has been renowned for its large estates. At the time of the New Domesday Survey in the 1870s, 50 per cent of the county was owned by the landed aristocracy, a proportion higher than any other English county (except for the diminutive Rutland) (Newby, 1987: 62). Contemporary data on private landownership is notoriously difficult to gather. One indicator of the extent of landed estates is the proportion of rented farm land. By the late 1980s, MAFF census statistics showed that wholly rented farm holdings occupied only 22 per cent of farmland in England; yet the proportion in Northumberland – at almost 45 per cent – was the highest of any English county. However, these figures underestimate the true extent of landed estate ownership because they exclude holdings that are partly rented and partly owner-occupied, holdings that are farmed in hand by the estates and areas of land subject to joint venture arrangements with farmers. We would estimate that large estates cover around a half of the county's territory.

The largest estate, which belongs to the Duke of Northumberland, comprises over 4,133 hectares acres, plus the mineral rights to 12,400 hectares. There are about a dozen other estates in the county with holdings in excess of 4,200 hectares and there are also about a hundred landowners who have more than 400 hectares. The sheer size of some of the estates has been a crucial factor in their survival as they are sufficiently big to weather shifts in the economic climate. Most estates, however, are not huge, but many of them have diverse sources of wealth. Indeed, the distinctive economy of the North East over the past two hundred years has helped to maintain and replenish the landownership structure of Northumberland.

The North East, and particularly Tyneside, led the way in the industrial revolution. The regional economy was based on the exploitation of coal, and Tyneside developed as an industrial heartland of heavy engineering and shipbuilding. The coalfields of the North East came to supply both Tyneside and the rest of Britain with its fuel. So landowners directly profited not only from increased agricultural returns (pushed up by the rising urban food demands of the burgeoning cities) but also from the extraction of coal and the development of the roads, railways and harbours to transport it. While by the mid-nineteenth century some members of the old aristocracy were among the most prominent of the coal owners, others leased out their coal bearing land and benefited from way leaves, royalties and railway rents. At the same time, minor landowners sitting on rich seams were able to build up their estates with the wealth that coal provided. Moreover, a successful mine owner would naturally aspire to join the ranks of rural landowners and to become a member of the country gentry.

Great houses and estates supported by coal wealth (the 'Coal Barons') came to occupy much of southern Northumberland and the Tyne valley. Subsequent generations of industrialists with wealth derived from engineering or shipbuilding had to look further afield. Some of the new industrial plutocrats acquired country estates in the north of the county. Thus, in a reinforcing flow of investment and economic development within the region, the coal fuelled the growth of heavy industry on Tyneside, while the new coal owners and industrialists used their profits to purchase estates in the surrounding countryside, and landed capital benefited from mining revenues, urban rents from the expanding towns and cities, and buoyant agricultural returns.

Most of these well-to-do families combined extensive landownership with a range of other interests. They were thus bound into a complex set of economic linkages. The Ridley family provide an interesting example of the blend between farming and non-agricultural activities that existed in the management of many estates in the North East. During the early eighteenth century the family became established as important landowners in south-east Northumberland after many generations as successful Newcastle merchants. Successive generations of Ridleys were not simply members of the landed gentry, but the proprietors of an estate of varied interests, all of which were sedulously cultivated:

> They certainly took care that their farms should be profitable, but they were quick to exploit all of the other opportunities for profit embodied in this substantial investment. Banking, coal-mining and the growth of the small port of Blyth were all Ridley interests which combined to make the family richer and more influential.
>
> (McCord, 1979: 36)

This pattern of activity has continued down to the present time. For instance, the development of Cramlington as a new town by Northumberland County Council mostly on land owned by the Ridley estate released large quantities of capital which were re-invested elsewhere.

Thus the diversity of the wealth of landed families in Northumberland has also been a major factor in ensuring their continuity. It has been bolstered by periodic injections of new capital throughout the twentieth century. After the Second World War when the Labour government nationalised the coal mines, the landowners received quite generous compensation for their mine holdings which in many cases were no longer profitable. The post-war farming boom also replenished the wealth of those estates that depended heavily on agricultural rents and incomes.

In short, the landowners continue to comprise a solid socio-economic formation at the core of Northumberland's rural economy and society. This formation sets the context not only for the configuration of development patterns in the region but also for the assertion of particular conventions. It controls access to land and therefore the types of new development opportunities that can arise. As we shall see, these property rights also allow it to shape the region's rural social structure. To illustrate the way in which today's landowners play these roles we outline below examples of both old and new landed estates.

The Duke's estate

Most notable among the landed estates in Northumberland is that of the Percy family – the Northumberland Dukedom. The estate is centred on Alnwick Castle, the Duke's home, which has been owned by the family since 1309. The father of the present Duke, whose long ducal career from 1940 to 1988 spanned the years of post-war agricultural prosperity, was very much in the tradition of the improving landowner and he concentrated on upgrading the estate's farming potential as well as expanding its forestry activities. The present Duke has a somewhat different outlook. He trained as a chartered surveyor and practised as a land agent before his accession. He therefore brings the perspective of a modern land agent to his role.

After completing his training, he was given the responsibility of managing the family's 1,240 hectare estate in Surrey. The estate was considered a liability because of its mixture of common land, relatively unproductive woodland, poor soils and public pressures. The Duke's father thought they should get rid of it. The son, though, set out to exploit its non-agricultural potential: 'we developed every little shed or outbuilding or garden conservatory into small businesses at a reasonable rent so that people who want to work in the countryside can do so and yet you are quite close to Dorking or Guildford'. The result was to turn around the income of the Surrey estate from about 70 per cent dependency on agricultural rents to about 20 per cent dependency. Having gained this experience, the present Duke speaks approvingly of 'modern estate managers and agents' as being 'a different breed to what they were twenty years ago. They have a much greater knowledge of urban and industrial needs and the broad spread of factors that are vital for a modern estate'.

Since his accession to the dukedom, his strategy has been to diversify the income streams of the family's Northumberland holdings too:

> our philosophy is to not have all our eggs in one basket, to spread out assets to try and get income from other sources – industrial, shops, housing – everything really. Our dependence on agriculture has to change because I cannot see the rental income from agriculture improving.

Of course, there is not the intensity of possibilities as in Surrey, but the vastness and geographical spread of the Northumberland holdings do present a range of opportunities. The core of the estate is in central Northumberland with large blocks in the North Tyne Valley and in the middle of Kielder Forest, but there are outlying parcels also in southern Northumberland including on the urban fringe of Tyneside and in the Tynedale commuter belt. Many whole villages are owned by the estate and a number of towns, including Alnwick and Rothbury, are set within it. The development opportunities include exploiting mineral resources and opencast coal mining, establishing light industrial and commercial units in favourable locations, building commuter housing on the edge of villages, conversion of redundant buildings to holiday accommodation and promotion of a multiplex cinema complex in south-east Northumberland. As a land agent in private practice explained:

> The Duke has been prepared, and this is especially so with the downturn of agricultural incomes, to diversify his business base and to look at alternative uses in particular of buildings surplus to requirements. The result has been to breathe new life into existing buildings and provide job opportunities for people living both on and off the estate. This is of vital importance as the labour force on farms becomes increasingly scaled down to the absolute bare minimum.

In many respects the Duke's outlook is that of a property developer, but he is conscious of limits to this analogy. First, the vast majority of the estate is and always will be in agriculture and forestry. Second, the estate aims to retain ownership of its diversified assets in order to enhance its long-term income flow, preferring to let commercial land or property on, say, a ninety-nine-year lease rather than to release it completely. Occasionally, small amounts of development land are sold off, perhaps around Alnwick or Tyneside. But when this is done the estate usually ploughs back most of the capital into buying more land, typically a small farm close to a village with development potential. Third, the estate takes a very long term view of its ownership of land. As the Duke puts it: 'it is engrained in us to take the long view . . . you have got to live with the consequences hopefully for hundreds of years'. Fourth, that view of the future stems from a strong appreciation of the historic continuity of the estate and the Percy family's identification with Northumberland.

The Duke clearly recognises that various historic obligations rest upon him, including the responsibility to ensure the long-term prosperity of the estate, to uphold its historic heritage, not least Alnwick Castle, and to conserve the rural environment. As he explained, when making development decisions

> it is always a fine balance. You have to look at it from a commercial point of view, and from the point of view that you are actually supporting everyone that works on the estate and supporting the heritage and all that.

Heritage conservation, although a considerable responsibility for the estate, brings in additional commercial opportunities. Alnwick Castle attracts a large number of visitors each year and the Castle is also regularly used as a film location.

The Duke's sense of responsibility for the local area extends to those people dependent on the estate: including the estate workers, the tenant farmers and the local communities in places like Alnwick. In the Duke's words: 'I do feel a strong link with the community'. He sees the estate's principal social role as an employer and explains the necessity for increasing the estate's income in terms of the need to keep so many people employed looking after the estate and the castle (the estate currently employs around 180 people). However, many more depend upon the estate indirectly for their livelihoods, such as the farm workers on tenanted farms or those employed in tourism in Alnwick. As the Duke also explained 'we are still one of the biggest employers in the area either directly or indirectly and I am very keen to increase all types of business to increase local employment'.

The fortunes of the locality and the Northumberland Estate are thus entwined. However, in discussing the relationship, the Duke is uncomfortable with the use of the term 'paternalism', preferring instead the image of the modern businessman who recognises his responsibilities to the local community. Nevertheless, he does acknowledge the interconnectedness of his business and the local community, explaining that

> the strategy of the estate, for many years, has been to try and develop the area in the best possible way that works both for us [the family] and the town. . . . Perhaps because the estate is so big and because of the history, there is a sort of paternal feeling – and I don't at all mean that patronisingly – I think it is a sort of general feeling within the community. It's a two-way thing and it is certainly felt by us as a responsibility.

We see here, then, the embedding of economic conventions in some rather traditional ideas about the rural community and the character of the locality as a whole. The Duke, like many estate landlords in Northumberland, endeavours to combine developmental, social and ecological aspirations.

This local embeddedness does not, however, imply that the Duke exercises direct political control over development. In fact, he explicitly rejects the political status that once seemed to be an integral part of rural landownership: 'it is anachronistic now to inherit leadership in a democratic society' or to assume 'a sort of almost sub-royal association with the county'. Unlike almost all of his predecessors he does not hold any political office either locally or nationally. In place of formal power and status hierarchies, the Duke – or at least his agents – must nowadays work through local planning and development networks to promote the interests of the estate. The estate's land agents are in frequent contact and discussions with local planners in the district and county councils, not only over the development of particular pieces of land but also over forward planning. As a result of its scale, the estate can plan its own land management and land development priorities several decades into the future and co-ordinate them closely, and from the earliest stages, with the preparation of statutory plans. It can also plan investment in major conservation, rural development and heritage projects in partnership with a range of local and national agencies in order to access public or European funding.

We thus see a long-standing and traditional economic institution (the estate) now entwined within local development networks in terms of planning and economic development. The gradual diversification of the business means that it plays a key role in regulating the emergence of any new economic networks in the area. However, while economic efficiency and making the best use of developmental opportunities are clearly important to the Duke, these conventions are set within a localistic context. Community, environment and tradition all come together within a repertoire of localistic evaluation. This repertoire provides strong continuity in the practice of paternalism in both the estate and the region as a whole.

The Joicey Estate

There are many medium-sized estates in the Northumberland countryside. One such is the Ford and Etal Estate in the Till valley in the remote far north of the county. It has been owned by the Joicey family since 1908. The family's wealth was originally based on engineering and coal mining. The estate encompasses some 6,500 hectares comprising 32 tenanted farms and around 250 dwellings. At the heart of the estate are the two villages of Ford and Etal. In the past, practically all of the villagers would have been employed on the estate itself (in forestry, gamekeeping, joinery and maintenance and shepherding) or in the estate's larger houses (as gardeners, grooms, cooks and other domestic staff). As employment opportunities on the estate have diminished, so local people have had to find other forms of work. Even so, about half of the adult residents of the two villages are either employees or retired employees of the estate.

Since the 1970s, a strategy of diversification has been vigorously pursued by the estate managers. A water-driven mill was restored as a tourist centre and this triggered the establishment of a range of other tourism-related activities. Redundant farm and estate buildings were redeveloped to house craft and tourism-related activities. The two villages of Ford and Etal have long attracted a small numbers of visitors. Each has an imposing castle and church. Ford school has murals painted in the mid-nineteenth century by the Marchioness of Waterford that depict Biblical scenes in Northumbrian settings with the local villagers as models. Etal has the only thatched pub in Northumberland, and its castle (now a romantic ruin) was involved in the Battle of Flodden. Following the appointment of a Commercial and Marketing Manager, who markets Ford and Etal using an 'umbrella' concept to cover the twenty-one different local attractions (including cafes, craft shops, castles and a light railway), visitor numbers have climbed to approximately 35,000 per year. Between them, these independently-run enterprises provide employment for around sixty people.[4]

Lord Joicey sees these developments as being in keeping with the outlook of a responsible landowner. In North Northumberland – 'an area dominated by landed estates' – it is 'the old estates that hold the cards in the creation and sustainability of rural employment'. Not only does this place responsibilities on the landowner, it also exposes them to charges of autocracy: 'I have even heard the word "dictator" used, and I can understand why. This is certainly the case for a larger landed estate, where the buildings or sites concerned are all under one ownership'. But the advantage of this ownership pattern, in his view, is that the estates are able to act decisively. The risk elsewhere is that

> where houses and land are owned individually, the potential for comment and possible opposition to any proposed scheme is perhaps greater. There is a corresponding need for more parties to be brought into the debate, more talk, more discussion, more diplomacy, all of which, in turn, might indicate a slower rate of progress in the creation and development of rural businesses.

Lord Joicey is keen to see the Ford and Etal estate develop in a 'balanced' and 'sustainable' way. In particular, he is concerned to ensure that his villages are places where people both live and work. He disdains the situation found 'in much of England [where] it is now commonplace for next-door neighbours in a village to work sixty miles apart from each other, and only seldom to meet'. In such circumstances, 'life at the workplace and life at home are now almost completely separate features, where once they were deeply interlinked. In north Northumberland, I would argue that they still are'. The consequence is, he believes, 'a more highly developed sense of community thinking prevailing in Northumberland compared to other areas of England'.

The estate therefore selects new tenants for its business premises with care and due regard to their suitability (i.e. their likely 'local embeddedness'). For instance,

> it is the estate's policy to ensure that the individual attractions complement, rather than compete with each other. Thus, for example, there is probably only room for one or two craft shops, and catering outlets. Too much competition for relatively few visitors is not worthwhile.

Lord Joicey is also concerned that the growth of tourism should not get out of hand:

> Ford and Etal must not be slaves to ever increasing visitor numbers. The villages remain places in which people live and work. To encourage and sustain a healthy village community with school, post office, village hall, etc., a village itself must maintain its own identity, and not become a museum.

Residents should not be hindered by tourist traffic, therefore visitors are deliberately channelled into a corridor between the two estate villages, with a few walks and riding routes radiating out from this. Lord Joicey is also of the view that

> visitors come to this part of Northumberland . . . for the peace, quiet and beauty it affords. Too many visitors in a sensitive rural area may be counter-productive. Ford and Etal Estates are concerned that too much promotion and development of purely tourist-orientated features could destroy the countryside and its attractions.

He draws a distinction between an employment strategy oriented for growth – i.e. more numbers employed – and one oriented towards sustainability – i.e. greater security for those currently employed:

> The modern world would of course like to have an instantly quantifiable growth factor to point to. For the rural communities of Northumberland the preference would be for stability and sustainability. In very rural areas, growth can bring its own problems.

We again see here the landlord executing a 'trade-off' between growth and preservation.

Not only does this attitude influence the type and scale of tourism facilities that the estate chooses to develop, it also shapes its promotional policy which no longer aims to increase visitor numbers in the peak season but to spread the season by targeting the type of visitor for whom seasonality is not an issue, thus favouring the extension of existing part-time employment rather than an increase in seasonal work. Since the mid-1980s, moreover, after experiencing a rapid turnover in its craft and tourism-related businesses and discovering that those firms that remained (e.g. woodworking and bespoke furniture) derived only a small proportion of their income from visitors, it has been the policy of the estate to offer redundant buildings to economic units whose principal activity lies outside tourism, and who have an independent, year-round market for their goods and services. Such businesses include spinning wheel manufacture, dyeing, architectural salvage, glass and can re-cycling and a bakery. While these may add to the tourism attractions – visitors are welcome to look round the workshops – the trade so generated is largely incidental. As Lord Joicey admits: 'The tenants of these buildings are experts in their own businesses. The estates are simply the landlord, providing the opportunity for the creation and development of small rural businesses'.

Again, in this case, economic conventions are mediated through a concern for the locality. On the one hand, the Estate is re-inventing itself, moving away from a concentrated focus on agricultural estate management to embrace new business functions. It derives its non-agricultural income from business ground rents, royalties from the use of the light railway and visits to Etal Castle, and some residential rental income from letting cottages and houses to the operators of these businesses. On the other hand, this diversification, while linking the estate into a range of economic networks, is kept within fairly strict limits so that the locality's rural character – in terms of community and environment – can be maintained. In part, the maintenance of this character is to ensure that economic activities can be sustained in their present shape (e.g. tourism). But it also derives from the landowner's interest in keeping rural traditions alive.

The Vinson estate

Our third example of a Northumberland landowner is Lord Vinson, who is a newcomer to the county. After National Service he joined a small plastics company. Having spotted the opportunities for plastic coating of metal baskets in dishwashers, he set up business in his own right. The company grew and prospered and in 1970 Vinson sold up, becoming a wealthy man while still only in his late thirties. He decided to settle with his family in Northumberland, having visited the county a few years earlier. The rural environment attracted him, partly because he is keen on country sports but also because, as the (fifth) son of a Kent farmer, he was attracted to the life of a 'gentleman farmer'. He wanted to live in a traditional, unspoilt area: 'This is a beautiful area, unscarred by modern development . . . I grew up in rural Kent on a farm, but the Kent countryside is under pressure. It has changed completely'. He wanted a social base from which to pursue his public and political interests. As he reflects, 'I made my money elsewhere. I was very fortunate. There are other things in life than making ever more money'.

He saw a farm for sale in *Country Life* and bought it in 1971. It is a 420-hectare hill farm at Roddam on the edge of the Cheviots with a fine country house, and that is where he lives. The following year he bought the Hetton Estate, a mixed arable/livestock farm of 1,250 hectares in North Northumberland, from the Co-operative Wholesale Society. Both of these large holdings are farmed in-hand and each is run by a general manager. He sees the quality of staff as a key aspect in the successful running of these enterprises: 'I need very good people so that I can do all the other things I do'.

Some thirty people are on the payroll at the Hetton Estate, some of them part-time. Lord Vinson refers to it as a 'self-sufficient agricultural business'. It is almost completely self-contained in terms of its labour needs. The main enterprises are dairying, arable crops, sheep and beef (the male Friesians from the dairy herd). As well as the farms, he has other business interests and is not dependent upon income from the estate. 'For most of the years it has more than paid its way. The 1970s, and the early 1990s, were good years. Currently I'm having to subsidise it.'

There are forty dwellings on the Hetton Estate (there were more, but he has sold off the houses in one small hamlet to outsiders). Most of the houses are let in-hand to employees and pensioners of the estate. The rest are rented out commercially. He used to let these as holiday cottages, but he realised it was better to have around-the-year tenants – the holiday season being too short in Northumberland. The only other diversification on the two farms is some sand extraction on the upland farm.

Although an in-migrant to the area, Lord Vinson has willingly taken on the mantle of stewardship of the traditional landowner: 'I prefer to see myself as the custodian rather than the owner of the estate' he comments. This has both environmental and social dimensions. Regarding the former, while he has sought to be an efficient and progressive farmer, he is also an enthusiastic conservationist. He has planted trees to enhance the look of the estate and has left unimproved some patches of old grassland along the river running through the estate which attract lots of snipe. The estate has been entered into the Countryside Stewardship Scheme which permits such practices as leaving uncultivated strips around arable fields. He sees his conservation work – for which he has won prizes – as being of wider civic value. Indeed, he has pushed through major changes to the footpath system on the Estate, including a lot of rationalisation and improved routes and an additional permanent bridle way. He is clearly pleased with these improvements.

However, the attraction of rural Northumberland is as much social as environmental. Lord Vinson himself is a devotee of country sports – shooting, fishing, horse-riding. There is a strong social side to these activities in the area:

> Hunting and shooting are an integral part of rural society in Northumberland. It's a factor in keeping the landed class here. . . . You might say that moving here I made a lifestyle choice, to be part of a traditional lifestyle. . . . Here country life still continues. This is a living agricultural community.

His sense of stewardship as a landowner includes also that community:

> There is a social side to landownership. You can't just lay people off when your returns are down . . . We also pay 10 per cent above the minimum. We want to pay a living wage. It does mean that we keep our people . . . we have a low turnover rate. They live rent and rate free with free water and free milk. Also we continue to house retired workers and widows. There's a cost involved, but it's one of our obligations as a landowner. It's expected of us. Other landowners do the same. In fact workers see it as part of their pension.

Yet, he also holds strongly neo-liberal political and economic views. He was a founder and financial backer of two very influential right-wing think-tanks, the Institute for Economic Affairs and the Centre for Social Policy. These developed much of the policy thinking of Thatcherism, and Vinson was made a Lord by Margaret Thatcher. However, he is not a free marketeer when it comes to agriculture:

> low agricultural commodity prices come about as a result of governments attempting to safeguard their food supplies and thus encouraging over-production. Nobody wants under-production and consequent starvation in the Western world! So it seems to me that unless agriculture gets some protection there is no alternative to either massive bankruptcies or the screwing down of wage levels in farming below the minimum wage – which would be politically unacceptable. Interference in agriculture has been with us since Roman times and is unlikely to go away. . . . The real problem is that food is too cheap.

This case differs a little from the previous two landowners: here, economic conventions are pushed into the background behind localistic, civic and environmental concerns. Having made his money elsewhere, Lord Vinson expresses these alternative repertoires in the management of his estate, which in his view lies 'outside' strict forms of economic calculation. The estate enshrines his desire to maintain rural traditions and practices and, in large part, his management of the estate is oriented towards these non-economic aspects.

The Davidson estate

Another new entrant to Northumberland landed society is Duncan Davidson, the owner of Persimmon Homes, a volume house-building company. Davidson first bought 190 hectares along with the stately home of Lilburn Tower in 1972 in North Northumberland to be his family home, with the intention of acquiring adjoining land as and when it became available. Over the past twenty years he has built up a sizeable new holding. In 1982, 1,860 hectares acres of neighbouring land were purchased from the Co-operative Wholesale Society, and during the 1980s and 1990s a further twenty-three hill and lowland farms were purchased to create the substantial current Lilburn Estate. The estate covers 11,158 hectares, extending south from the Scottish Border. Most of it is farmed in-hand: just 1,300 hectares are let to five tenant farms. The farming enterprises are mainly livestock-based: there are 820 hectares of forestry, 1,240 hectares of arable land and the rest is for grazing, including 5,785 hectares of hill land. As farms have been acquired, buildings and fencing have been repaired and modernised, and the estate now has a workforce of 60 people, including a general estate manager, an assistant manager, a farm manager and a financial manager. A further 15-20 contract workers are employed from the local town of Wooler and the surrounding areas to undertake all the masonry, joinery, plumbing and electrical work.

With a clear sense of purpose Davidson has transformed his accumulated holdings into one of the leading shooting estates in the country. A 100-hectare formal park has been established out of former arable land, complete with its own lake and folly. A wetland area has been formed to encourage wading birds to over-winter and breed. On the low ground of the estate a high-quality pheasant and partridge shoot has been established. On the higher ground, as hill farms have come into its possession, sheep numbers have been systematically reduced to turn grazed hill land into a heather-clad grouse moor which in a good year now yields 1,250 brace. The estate is playing a leading part in an ambitious project to re-introduce the Black Grouse to the Cheviot Hills. To complete the picture of an estate offering the range of traditional pursuits, a fish pass has been constructed to a weir to allow sea trout and salmon to move up to the head waters of the Lilburn which runs through the estate.

What others might see as a grandiose exercise in conspicuous consumption, Davidson himself regards as a demanding and compelling project to establish a model working estate. While made possible 'by being fortunate enough to have the funds from non-farming businesses, with which to purchase the land', he is motivated by respect for the land and its people.[5] In his own words, 'the farmland, woodlands, parks, moorlands and gardens reflect the high respect we have for the land, the landscape and the people who live and work upon it'. This comment implies an intention to embed the economic activity taking place on the estate (notably the shooting) in the traditional social and environmental features of the locale. The landscape, the workforce and the surrounding community are all seen as expressions of a relationship in which the estate and the rural character of the area are closely aligned. In this respect, the Davidson estate reflects the wider set of paternalistic conventions that are deeply inscribed in the social and material fabric of rural Northumberland.

Paternalism and clientelism: large landowners and the state

These cases indicate how the large landed estates establish hierarchies of conventions in the countryside. The estates are managed as commercial concerns but management practices are sensitive to the assumed needs and traditions of the locality's population and environment. The landlords' (some of whom come from outside the long aristocratic lineages that run through Northumberland society) reflexively orchestrate an array of conventions in line with their own conceptions of estate tradition. Thus, the estates constitute 'orders of worth' in which commercial imperatives are embedded firmly within an assemblage of heterogeneous resources, arranged in ways that maintain stability from the past into the future. The economic, social and environmental values embodied in the estates are consolidated in a self-reproducing social formation. Its values, though based in the past, are oriented to the future. As one land agent commented: 'All landowners have one overriding consideration and this is to preserve their assets for the benefit of forthcoming generations'.

The region's rural landscape thus reflects the landlords' paternalistic aspirations. Indeed, the significance of the estates is such in Northumberland that it is hard to avoid the symbolism of the local landowners and their ancestors. Castles abound, dominating valleys and coasts. Other aristocratic elements make up the rural scene, including historic parklands, the uniform architecture and layout of the estate farms and villages, the many examples of local benefaction such as churches, fountains and village halls, and the formal monuments to past nobility. This ancestral landscape is an important legacy and defining context for established owners. It is also a key attraction for those who want to buy into traditional society, and who seem ready to accept the conventions of paternalistic stewardship.

The values of landed society are also projected beyond the county and into the public realm. The lordly relics that now constitute Northumberland's rich 'heritage' are extensively drawn upon in constructing the aesthetic qualities of the county and in projecting the area as an attractive and distinctive location or destination. The Northumbria Tourist Board boasts that Northumberland has more castles open to the public than any other county. More controversially, the lifestyle of the Northumberland 'countryman' – particularly the pursuit of country sports – is also widely projected. Many of the estates derive significant income from the commercialisation of shooting and fishing, and there is strong backing from the county's landowners for the national campaign to protect hunting and other country sports from legal prohibition.

However, the practices of the large private landowners are not the only decisive influences in the countryside for we can also observe the interplay between paternalism and what we have called 'clientelism' (Marsden *et al.,* 1993). This latter category was defined in *Constructing the countryside* as those localities in which state agencies play a major direct role in land management and rural planning. As well as the large private estates in Northumberland, the county also contains a number of institutional landowners – the Ministry of Defence (MoD), the Forestry Commission, the Northumberland National Park Authority, the former regional water authority, the now privatised National Coal Board (NCB), English Heritage and the National Trust. This extensive institutional ownership in Northumberland occurred largely in the middle period of the last century when private landownership generally was in retreat. In a series of events, large tracts of land in Northumberland were transferred from private estates into public control.

One early example took place in 1911 when 7,600 hectares of the Redesdale Estate were bought by the War Office as an artillery range for the newly formed Territorial Army. It is said that the transfer came about after Winston Churchill as Minister of Defence had been out shooting with Lord Redesdale and had said that he needed to try out a new weapon (the Howitzer) on this sort of land. Since then, the military's land-holdings in the county have grown and currently amount to 23,120 hectares of freehold and leasehold land, including thirty-one tenanted farms. The bulk of this estate is made up of the Otterburn Training Area, currently used for training troops in the use of heavy artillery, and which comprises almost a quarter of the Northumberland National Park (Woodward, 1998).

In a similar manner, the Forestry Commission has built up its holdings in Northumberland. In 1932, the Commission was able to acquire 19,000 hectares of land in the North Tyne catchment from the Duke of Northumberland's estate. The sale was forced because the estate had to raise death duties following the deaths of two Dukes in quick succession. The transfer included the Duke's former hunting lodge, Kielder Castle, and the land acquired became the core of the Commission's new Kielder Forest. Land continued to be acquired, until by the early 1970s there were over 50,000 hectares of forest plantations.

Through the transfers of such large blocks of land, the Northumberland landscape became 'carved up' between institutional owners and the local private estates. For some of these institutional landowning interests – such as the National Coal Board (NCB), the Forestry Commission and the Cooperative Wholesale Society – their approach to land management and rural development was, in part, driven by socialistic planning, in deliberate opposition to the private estates. Those on the Left of British politics regarded the large estates in a county such as Northumberland as bastions of reaction, with a stranglehold over the economic development of the county's physical resources and the social improvement of local people. Thus, the Forestry Commission, as well as taking over and planting land, developed whole new villages for forestry workers. Clientelistic landowners therefore competed with the more traditional landowners.[6]

Initially, there was mutual antagonism between some of the new public sector landowners and the private estates. Yet, the institutional landowners have gradually developed a *modus vivendi* with the private estates and in certain respects have mimicked the latter's paternalistic social relations. To an extent, the institutional owners operating in the countryside have had to 'fit in' with the existing social formation. For example, they have always taken their lead in rental practices from the private owners, thus perpetuating a longstanding practice whereby the Northumberland estate established the going rate for agricultural rents and wages in the county. Simply as neighbours, the institutional landowners have had to conform to established conventions.[7] Moreover, public bodies that own land but also operate wider promotional and regulatory functions – such as the Northumberland National Park Authority, the Forestry Commision, English Nature, the Environment Agency and English Heritage – depend to a considerable extent on the active co-operation of the large landowners to achieve their objectives. Usually, the bodies advising or steering these agencies include prominent local landowners among their members.

In general, the 'clientelistic' (public) bodies and 'paternalistic' (private) landlords hold a common understanding of the overall countryside context and their obligations and interests within it. All are members of the Country Landowners' Association; moreover, the land agents, who make many of the crucial management and investment decisions and are in overall operational control, form a unified professional group within the Royal Institution of Chartered Surveyors. Significantly, while both these representative bodies straddle the private and public sectors, they are deeply imbued with the ethos of private ownership and the need to embed their estates within the overall regional socio-economic context. In other words, all the landowners subscribe to a common set of localistic, economic and environmental conventions.

These commonly-held conventions, which link the various landowners in an informal network, were nowhere more apparent than in the Otterburn Public Inquiry held in 1997. The MoD applied for planning permission to develop a series of new gun spurs, widen roads on the training range and develop some buildings in order to accommodate new training requirements, all on the Ministry's existing holdings within the Northumberland National Park. These changes, which were associated with the use of Multiple Rocket Launch Systems (made famous during the 1991 Gulf War) and AS-90 weaponry, represented an intensification of the use of the range for military training and were met with opposition from amenity interests and the two local planning authorities – Northumberland County Council and the National Park Authority (see Woodward, 1998).

The Otterburn debate was, in part, a debate about paternalism. The military claimed that, in addition to its responsibilities for the defence of the realm, it was also concerned to protect the local area, and not just the wildlife. The MoD made much of its own contribution to the local economy and to environmental management. Rural stewardship conventions were drawn upon strongly before and during the public inquiry as the military argued that it had strong regard for the local community and its future. The MoD's rhetoric of protecting local people and jobs found echoes in the formation of a local action group calling itself the Association of Rural Communities, which *supported* the case for military expansion and attacked the National Park Authority for having challenged the military's plans.

Thus, through its practices and representations, the MoD helps perpetuate the paternalistic countryside. Of course, its very presence in the National Park also discourages the middle-class gentrification of the countryside, ensuring that the Park does not become either the contested countryside or the preserved countryside, but maintains a traditional rural society. The Northumberland National Park actually has the smallest residential population of any of Britain's National Parks, totalling only some 2,000 people. Significantly, it is the only English National Park not to have its own protection society. A counterpart to the predominance of paternalism is therefore the weakness of the middle class in the Northumberland countryside.

Preservationism in the paternalistic countryside

As we indicated above, counterurbanisation is not well advanced in this region and there is no ready constituency to aid the construction of preservationist networks. This is well illustrated by the difficulties that the CPRE has faced in establishing itself in the county. The CPRE has a fairly low public profile in Northumberland. For many years it relied on the Northumberland and Newcastle Society (NNS), a traditional county-cum-civic society which dates back to 1924. However, following a failed attempt to incorporate the NNS into the CPRE's national structure, a Northumberland branch of the CPRE was established in 1993. It started with 180 members but had built up to just 360 by the end of 2000. This membership is largely suburban or outer-suburban, from in and around the Tyneside conurbation. The CPRE has little presence in the more rural parts of Northumberland.

In fact, the CPRE's existence in the region relies on the continued support of the national CPRE office and its paid staff. Over several years, CPRE headquarters has funded an officer stationed in the North East to assist in the formulation of regional policy. A Branch Development Officer, whose specific role is to assist in establishing CPRE branches for Northumberland and Durham, is also based in the region. This central support has allowed the Northumberland branch to have an involvement in planning that, as one county planner observed, is 'disproportionate' to its local membership base. The CPRE is able to make regular representations to reviews of county and district plans, with housing and green belt issues being particular foci of attention.[8]

However, a consequence of this central direction is that the CPRE's involvement in local planning issues is 'erratic', according to one planning official who noted that they 'don't tend to get terribly involved' in contesting planning applications on a regular basis. This official believed that the haphazard nature of the CPRE's campaigns in Northumberland arose because they had taken the decision to 'concentrate their limited resources at the regional level on regional strategic matters rather than on individual applications'.[9] Moreover, the CPRE's involvement in environmental issues does not extend beyond planning and housing. The Director of Northumberland Community Council commented that the CPRE does not 'figure in county politics'. The Head of Economic Development at the County Council said he did not encounter the CPRE, although he was aware that they existed within the county. A prominent councillor, and leading member of the Liberal Democrats, could not recall having to deal with the CPRE in any of the positions he had occupied in the county, including membership of the Planning Committee, Chair of the Farming and Wildlife Advisory Group, trustee of the Northumberland Wildlife Trust and Chair of the county NFU.

The weakness of the local branch means that the CPRE comes to be associated with the national and regional offices. While this heightens its standing with policy professionals – who tend to agree that the representations coming out of these offices are of a high quality (and do not see them contradicted by more parochial branch outputs) – it adds to the perception that the CPRE is an organisation rooted 'elsewhere'. This perception is, in turn, reinforced by the selective identification of key issues by the national and regional offices: issues are thought to be of interest because they reflect *national* campaigns, rather than the most pressing concerns of people in Northumberland. In short, it is assumed by many in the region that the CPRE represents a view of the countryside that only makes sense in the 'south' (i.e. in the 'preserved' countryside). An economic development professional in the region who had previously worked in the Cotswolds put the point in the following way:

> where CPRE is stronger, it is a community of commuters. . . . And there is a correlation between commuters and NIMBYs: 'back to my darling village where nothing must change and we get very upset when cows actually crap down the road. It puts marks on our Porches'.

In contrast he said, 'Northumberland is not a county of commuters . . . and we have got a lot more countryside here'. In his view, the dominant perspective in Northumberland is of a 'developmental' countryside, a countryside in which economic and social, rather than environmental, conventions prevail.

This weakness in the CPRE is explicitly linked to counterurbanisation and development pressure. A leading figure in the NNS explained:

> There isn't much of a new middle class in Northumberland. Okay, there is up in Tynedale and up towards Alnwick, but my part of rural Northumberland is essentially the big estates with the landowner and the tenant . . . because there is a high level of tenancy, whether farmer, cottager or householder, there would be a certain risk in sticking one's neck above the parapet and saying 'this kind of development isn't welcome'. But also there is very little development going on.

From these comments it would appear that the prevailing view in Northumberland is that development is something to be welcomed not feared. The CPRE's perspective therefore sits uneasily in a political culture which prioritises jobs and development. For instance, a policy officer for the Government Office for the North East claimed that:

> The vast majority of people [in the North East] actually want development X. There may be opposition from CPRE. But, by and large, CPRE are on their own, whereas if you are in the South East they are probably much more reflecting the general view and in tune with public opinion down there. Up here, you often get the situation where the local authorities, all the economic bodies, the regional development agency and everybody else are saying 'we want X'.

The preservationist stance taken by the CPRE stance is seen as inappropriate. On those occasions when the group is active it is often accused of 'importing' protectionist views from the south of England, where a strong preservationist social formation exists, into the north, where both landlords and workers seemingly support new economic development.

Conclusion

The legitimacy of rural landownership in Northumberland rests on a particular set of conventions associated with landscape, nature conservation, the maintenance of a diverse and vibrant rural economy, and the maintenance of stable rural communities. These conventions are arranged in ways that embed the local economies of the estates within a rich array of social and environmental relations. And these relations are not just confined within the estates themselves – they extend out to influence the character of rural areas in the region as a whole.

The landowners comprise a mixed group: some have been born into the locale and have adopted the landlord's role as a birthright; others have made their money elsewhere and have taken on the mantle of paternalistic stewardship. Yet, this distinction blurs in practice as the 'old' landlords adopt modern estate management techniques and as the 'new' landlords adopt traditional attitudes. In such ways, inherited forms of paternalism are (reflexively) re-shaped in the modern era. For instance, we find that although the landlords aim to act in keeping with long-standing traditions, a key feature of paternalism, the assumption of political leadership, is now absent from the countryside of Northumberland. Within the space of a generation, Northumberland's larger landowners have largely withdrawn from local politics.[10] As one smaller landowner said of the larger estate owners, 'they tend to keep themselves to themselves'.

Yet, although the landowners now tend to eschew the political leadership role that was automatically assumed by their forebears, as we have emphasised, this has not diminished their local influence. Particular estates occupy strategic positions geographically, in relation to, for example, the growth of certain settlements, access to key natural or mineral resources, the suburban growth of the conurbations and so on. They are thus well placed to control specific development processes. By only releasing small amounts of land for very selective forms of development, the landowners have prevented the suburbanisation of the countryside, thereby helping to exclude the counterurbanising middle class. Moreover, they have maintained the countryside as largely a 'developmental' rather than as a 'preserved' space. Thus, the sentiments of preservationism expressed by counterurbanisers seem largely out of keeping with the prevailing socio-political context (except, perhaps, where middle-class aesthetic concerns coalesce with the environmental and social aspects of paternalism).

The absence of a rural middle class in Northumberland is evident from the weak preservationist networks operating in the region. As we outlined in relation to the CPRE, the assertion of preservationist conventions is made doubly problematic by the absence of a sympathetic social constituency and a widespread adherence to conventions of development and economic growth among influential rural residents, politicians and policy-makers. Moreover, the power of the estates reduces the significance of political and planning networks. Through their extensive and consolidated ownership of land, the estates hold what amounts to almost a local monopoly over development opportunities. Thus the development strategies of the estates are far more significant than the activities of local political and planning networks. The estate management and investment decisions of the larger landowners, both public and private, carry more weight in defining local development practices than those of the local authority. The latter is therefore re-defined as a negotiator with the estates rather than a full-scale regulator of private sector development in the countryside.

In short, the reflexive middle-class networks that are active in the preserved and contested countrysides are much less significant in the 'paternalistic countryside'. They are constrained by the continuation of a quite traditional social formation established around the large estates and the private and public landlords. While many of these estates are modernising their business practices and diversifying the economic activities in which they are engaged, any such innovations are set *within* the local community and the local environment. In fact, economic innovation is usually undertaken in order to *maintain* the traditional character of these communities and these environments and is therefore evaluated in terms of its continuity with past practices. In this fashion, paternalistic landowners can be seen as pursuing a form of 'aesthetic reflexivity' but here the goal is to maintain rural traditions in a changing economic, political and social context. These traditions still define the distinctive character of rural Northumberland, but are now asserted in ways that combine long-standing paternalistic concerns with more modern economic and environmental sensibilities. This new mix of conventions underpins the contemporary expression of paternalism in the countryside.

7 The differentiated polity

Introduction

In the introduction to this volume we argued that two main narratives shape the English countryside: a narrative of 'pastoralism' that sees rural areas as properly lying outside industrial or capitalist modes of development; and a 'modernist' narrative that wishes to incorporate the countryside within rationally administered and progressive processes of change. The first sees the countryside as a pre-modern space; the second sees the countryside as requiring integration into modernity and its institutions. We proposed that the relationship between these two narratives varies according to the structural and spatial context.

In Chapters 2 and 3 we showed that the structures governing rural space are changing at the present time. In particular, new industrial agglomerations are forming which seem to span both urban and rural areas. Thus, in economic terms the countryside is brought within contemporary processes of change (although its incorporation is often taking place with an acute sensitivity to certain rural characteristics, notably environmental assets). In social terms, the countryside is also being incorporated within the general social formations that span urban and rural areas. However, here the incorporation is being limited by the desire on the part of middle-class political actors to preserve the countryside in line with the pastoral impulse. Thus, counterurbanisation both incorporates the countryside into modern processees of change and simultaneously sets this space apart.

In Chapter 4 we described the 'preserved' countryside through the case study of Buckinghamshire in the South East region. Here we found that high levels of counterurbanisation have generated a heightened sensitivity towards any development threats to the countryside environment. Such sensitivity is expressed in a hierarchy of conventions in which economic and development concerns are routinely subordinated to aesthetic, localistic, communal and environmental considerations. This conventions hierarchy is held in place by an environmental network that combines county and district councillors, amenity groups and local residents. The network is active within the planning domain in order to influence both plan policies and the follow-on decisions over development proposals. It is confronted by developers seeking to meet economic demands associated with a buoyant regional economy. The development network utilises national and regional policies in order to overcome localistic conventions and their associated constraints. Nevertheless, the outcome of the planning process, in which the differing networks confront each other, is a form of *containment* that steers development away from the countryside and into urban areas. Thus, a protected countryside sits cheek by jowl with rapidly growing towns.

In Chapter 5 we examined the 'contested' countryside in Devon. This countryside type bears some resemblance to the preserved countryside but here the contest between environmental networks and development networks unfolds around more localised issues. The development network, which encompasses local councillors, small businesses and some local residents, such as farm families, incorporates development concerns within a rural development repertoire. The environmental network regards this form of localism as inherently parochial, and believes that it leads to a compromising of valued natural environments. The environmental network thus utilises national and regional planning policy criteria in an effort to 'professionalise' the local decision-making process. Here, two convention hierarchies – one associated with local rural development, the other with protection of the environment – come into conflict. The contested nature of this countryside arises from the fact that neither of the networks can achieve overall superiority. However, we also found some scope for a mutual accommodation between the two views in their shared concern for the distinctively *rural* characteristics of the region. While this accommodation is at an early stage, it illustrates how new conventions hierarchies might be tailored to specific regional contexts.

In Chapter 6 we considered the paternalistic countryside of Northumberland. Landed estates are the dominant feature of this rural area, and landlords operate their estates as commercial businesses. However, economic considerations are refracted through the landlords' own aesthetic and localistic concerns. The estate businesses are embedded in the community and in the local rural environment, and are run in line with the tradition of paternalism, with the landlords reflexively stressing stewardship and continuity with the past. The estates are set within a context of limited counterurbanisation and the weak preservationist networks offer little real challenge to the practice of paternalism. As a consequence, the estates – along with associated artefacts such as castles, landscapes, villages and so on – give rural areas in Northumberland their symbolic character and ensure that change only takes place incrementally.

The three areas differ in important respects. The countryside in Buckinghamshire is decisively shaped by preservationism and urban containment. In Devon it is marked by a fissure between environmental and developmental networks. In Northumberland, it is stabilised by a long-standing social formation that is evident in the structure of land holding and in the rural landscape. In all these areas we see complex interactions between international, national, regional and local processes. Connections between the region and local rural areas are established by networks – primarily associated with economic, political, social and aesthetic structures and practices – and these generate conventions that configure development activities in the countryside. The shape of the differing rural localities described in the last three chapters can thus be attributed to the constellations of networks and conventions at play and the way these networks wrap themselves around the land development process.

It is clear that, at present, the differentiated countryside is poised between the desire to set rural areas apart from contemporary forms of development and the desire to see them change more in line with general patterns of modernisation. In the separate areas the constellations of networks ensure that very differing outcomes 'on the ground' are achieved. In some areas the countryside is preserved in line with environmental and localistic considerations; in other areas it is developed in line with economic conventions; in yet other areas some combination of the two determines local outcomes.

However, the structural context in which the narratives of preservationism and modernisation play themselves out is changing, as we saw in Chapters 2 and 3. At the present time, change is particularly rapid in the sphere of policy as government seeks to re-negotiate its activities and responsibilities in rural areas. Of particular importance, as we noted in Chapter 2, is the on-going shift from a national to a more regionalised structure of rural policy making. It therefore seems appropriate to consider in a little more detail the impact of regionalisation in the policy arena. In particular, we need to assess how far the introduction of a regionalised structure will alter the balance between the two narratives of pastoralism and modernism. In other words, will the regionalised state work to exclude rural areas from modern processes of development on environmental grounds or will it act to integrate rural areas more fully into general processes of change?

In this chapter we attempt to answer this question by considering prominent tendencies in the new regional structures. We first assess the modernising impulse as it is revealed in the evolving work of the Regional Development Agencies (RDAs). These agencies have been charged with bringing rural economic change more closely into line with general patterns of change. The RDAs thus seek to modernise the rural economy. Second, we turn to examine the new regional planning bodies. In many respects these bodies act as a counterpoint to the RDAs for they aim to ensure some amount of rural preservationism is achieved. We therefore see the two narratives at play in current policy approaches.

This chapter also examines some of the ways in which a more regionalised approach to rural policy may alter the constellations of networks and their associated hierarchies of conventions. In particular, we consider whether new policy structures will strengthen or weaken containment approaches in the preserved countryside, whether they will tip the balance towards either developmentalism or environmentalism in the contested countryside and whether they will confirm or diminish the centrality of the landed estates in the paternalistic countryside. In other words, we consider the types of differentiation that are likely to flow from the new structures of regional governance.

Recent trends in the regionalisation of rural policy

The modernising impulse

A new round of policy modernisation has been underway since 1997. We choose this date not just because it marks the election of the Labour government but also because it encompasses unprecedented reform in rural development and agricultural policy. At the European level, early 1999 saw the agreement of the Agenda 2000 reforms which established the new Rural Development Regulation as the so-called 'second pillar' of the CAP (discussed in Chapter 2).

Domestically, rural policy has also been reviewed in ways that have implications for the regionalisation of the CAP. For example, in 1999 the Prime Minister's Performance and Innovation Unit in the Cabinet Office proposed a strengthening of the regional dimension in rural development policy (PIU, 1999), a proposal which informed the preparation of the Rural White Paper for England published in November 2000 (DETR and MAFF, 2000). The PIU endorsed the argument that there should be a strong regional component to the implementation of the Rural Development Regulation in England. The Regulation requires member states to draw up Rural Development Plans for the period 1999–2006 containing agri-environmental schemes and other measures supporting rural economic development and farm diversification. The PIU report argued that national and regional partners in rural development should work together to 'generate integrated regional strategies for agriculture and rural development in the English regions' (PIU, 1999: 107). It also reviewed the structures and machinery of government for rural policy at the regional level and argued in favour of moving the policy-related aspects of MAFF's regional work into the Government Offices for the Regions, a proposal that was adopted by the government in 2000. Modernisation thus quickly came to mean regionalisation.

The government then decided that the Rural Development Regulation would be implemented in England by means of a national plan – the England Rural Development Plan (there were separate plans for Scotland, Wales and Northern Ireland). The English Rural Development Plan contains nine regional chapters which set out the rural development challenges from a regional perspective. Although the Plan was produced to a tight time-scale, it effectively represents the first attempt to manage strategically the roles of the agricultural and land development industries in each of the English regions. Thus, the Plan's regional chapters signal a first and important step in the regionalisation of a key element of the CAP, one that is likely to become increasingly important in the years ahead as further rounds of reform take place (Lowe and Ward, 1998). Moreover, regional consultations on the Plan brought together a range of stakeholders involved in agricultural and rural development policy within each region. The introduction of the Rural Development Regulation began a process of consolidation of the rural networks to be found in regional contexts.

Further impetus was given to the regionalisation of policy with the publication of a Rural White Paper for England in November 2000 (DETR and MAFF, 2000). The White Paper proposed a system of regional 'rural sounding boards' in order to facilitate participation in regional rural policy debates on the part of key stakeholder groups. This proposal was stalled by the outbreak of Foot and Mouth Disease in 2001, but in January 2002 the government announced the establishment of a Regional Rural Affairs Forum in each of the English regions outside London (DEFRA, 2002c). Each Forum has a seat on a national rural affairs body for England which is to act as a sounding board to Ministers in the new

Department of the Environment, Food and Rural Affairs. The Regional Rural Affairs fora are likely to develop important roles in liaising with new regional institutions, such as the Regional Chambers and Regional Assemblies, in order to increase dialogue between regional actors, enhance the regional sensitivity of national programmes and spending commitments and facilitate the emergence of regionally distinctive policies. They will thus help to consolidate networks oriented to economic development in the regions.

These various initiatives clearly aim at promoting 'institutional thickness' at the regional level in the sphere of rural development (Amin and Thrift, 1994) by incorporating rural areas into regional partnerships and networks. The rural is not only being regionalised thereby but is also being encompassed within a policy discourse that prioritises economic conventions of growth, competitiveness, efficiency and development. In short, the regionalisation process is aimed at the modernisation of the rural economy.

This is nowhere clearer than in the activities of the RDAs which are charged with developing a strategy for the economic development of their respective English regions. In stipulating how they should take account of rural issues, government guidance directed the RDAs:

> to ensure that rural areas benefit from and contribute to the development of the region as a whole, taking account of the linkages between town and country in transport, employment, recreation, the provision of food, fibre and fuel, and a recognition that the economy of rural areas has much to contribute to the competitiveness of their regions.
>
> (DETR, 1998c: 6)

The guidance set out the key features of rural economic activities and rural disadvantage, and made suggestions on how to tackle rural regeneration and deprivation. Previous experience, it explained, suggested that such efforts are more likely to succeed if they 'encompass measures to tackle economic, social and environmental issues in an integrated way' (ibid.: 40). Rural regeneration ought to encourage co-operation between different organisations and sectors, including the voluntary and community sector, and should support initiatives that take a long-term view.

Again, we see an emphasis on regional capacity building and rural network construction, with the overall aim of incorporating the rural into more general processes of economic modernisation. Rural issues and concerns thus figured in the Regional Economic Strategies that all the RDAs had to prepare. However, this was novel territory for economic planners and the business and public sector interests that dominated the RDA boards. The rural tended to be covered in a formulaic manner and with so many competing claims on them, it was uncertain what priority RDAs would accord to rural development.

The 2001 Foot and Mouth epidemic and the havoc it generated proved a catalyst for a fundamental re-assessment of rural development issues. It challenged many organisations, not least the RDAs, requiring of them a more active involvement in, and better understanding of, their rural areas. The epidemic, which began in February 2001, triggered a crisis not only in the agricultural sector but also in the wider rural economy. In a desperate effort to contain the disease, National Park Authorities declared their parks to be out of bounds, visitor attractions were shut and rural organisations cancelled events. It was some weeks before it was recognised that the likely losses faced by the farming industry as a result of Foot and Mouth would be dwarfed by the impacts of 'closing' the countryside on other rural businesses dependent on leisure, tourism and access to rural areas. While these impacts were felt widely across the UK (including in areas without any disease cases), they were locally and regionally differentiated in a way that reflected, indeed brutally revealed, the diversity of the contemporary countryside. A government that was overstretched centrally in fighting the disease therefore handed over the task of spearheading rural recovery to the RDAs. In a critical decision, the RDAs were asked to manage a Business Recovery Fund – to assist affected businesses – which cast the RDAs in a leading role in the response to Foot and Mouth disease.

We can speculate that the Foot and Mouth crisis has further shifted the policy emphasis from a focus on agriculture to a concern for the gamut of economic activities conducted in rural areas. As a consequence of this shift, the RDAs look set to play a leading role in modernising the rural economy. In so doing, it is likely that they will integrate rural economies into regional economies. The RDA's will be less concerned with urban–rural distinctions than with the economic buoyancy of their respective regions. Further economic differentiation looks set of follow.

The emphasis now being placed on RDAs in the context of rural policy holds the potential not only to strengthen rural economic capacities at the local level but also to generate greater interaction between differing network types in rural regions. In the long run, the RDAs are likely to develop distinctive strategies and approaches tailored to the requirements of their respective regions. The growing significance of such strategies implies that instead of rural development being an adjunct to, or separate from, agricultural policy, the latter will become an integral component of the former. Agricultural and economic networks may thus become more closely intertwined. And in this intertwining we can perhaps see the further integration of the rural economy into the more general economy.

The pastoral impulse

While the establishment of the RDAs has undoubtedly strengthened the assertion of economic conventions in rural regions, we should not assume that the economic agenda is now set to dominate the differentiated countryside. Environmental conventions and networks, as we noted in Chapter 4, have also gained from recent changes to the policy structure, notably in the planning arena. Although the regional level has traditionally been the weakest tier of planning in the UK, its status has grown since 1997.

The new Labour government proposed that regional planning should be seen not just as a tier in the policy hierarchy but as an important means of achieving horizontal co-ordination between regional 'stakeholders'. An early consultation paper, *The future of regional planning guidance* (DETR, 1998a), proposed improving the arrangements for co-ordination of land use, transport and economic development planning at the regional level. It argued that regional planning had previously been limited in its effectiveness, through an overly narrow focus on land use issues, a lack of attention to environmental objectives and a lack of transparency in the planning process. A new approach sought to ensure that the planning system would become more responsive to regional demands and sensibilities, with regional planning bodies assuming more responsibility in deciding how development could be accommodated. As the new draft PPG 11 (Regional Planning) put it: 'The long-term objective of RPG should be to develop into a comprehensive spatial strategy for the region i.e. to set out the range of public policies that will manage the future distribution of activities within the region' (DETR 1999: 4). In short, regional planning should reflect the needs of regional and sub-regional territories, including rural areas.

Thus, regional planning bodies were required to tailor planning policies to their rural areas. In doing so, they had to take on board the particular economic and social problems faced by rural areas and formulate strategies to combat such problems. The guidance places great emphasis on the establishment of working relationships with regional stakeholders. PPG 11 (in its final form – DETR, 2000b: 42) says:

> In developing advice on rural development, regional planning bodies will need to involve a wide range of organisations including the Government Offices, RDAs, the Countryside Agency, National Park and Broads Authorities, Association of National Parks Authorities, Association of Areas of Outstanding Natural Beauty, English Nature, English Heritage, Environment Agency, Forestry Commission, MAFF [now DEFRA], AONB Joint Advisory Committees/Conservation Boards, the Regional Tourist Boards and the Rural Community Councils. They should take account of work undertaken by these bodies, for example, the programmes of the statutory countryside bodies, MAFF's England Rural Development Programme and related programmes, the forestry strategy for England and agri-environment and related programmes.

Notwithstanding potential difficulties in co-ordinating such a diverse range of participants, there is a clear emphasis on regional network building in the policy statement. Rural planning policy is to be formulated and implemented by a diverse range of stakeholders working together at the regional tier.

In some respects, the regionalisation of planning policy might be seen as a continuation of the modernising impulse for it seems to imply that rural areas should be brought more fully into the gamut of policies operating across regional space. For instance, the guidance emphasises that sustainability approaches have a particular relevance to rural areas and must lie at the heart of rural planning policy at the regional level. Rural areas should therefore be incorporated within 'sustainable development frameworks' which would identify the key social, economic, environmental, and resource issues and the interrelationships between them (DETR, 2000d). Such frameworks appear to link patterns of development in rural areas and urban areas.

The construction of these sustainable development frameworks might be interpreted as a form of 'ecological modernisation' at the regional level. However, as we saw in Chapter 4, the interpretation of 'sustainability' in planning is strongly influenced by preservationist thinking. The preservationist agenda is being raised especially forcefully in relation to housing, notably the perceived requirement to limit the amount of new houses being built in the countryside. The influence of the preservationist perspective became especially clear following the publication of PPG 3 (Housing) in 2000 (DETR, 2000a). This PPG states that in formulating their housing policies, regional planning bodies should assess 'both the need for housing and the capacity of the area to accommodate it' (ibid.: 20). It is emphasised that regional planning authorities must take the latest household projections into account but these must be balanced against the 'capacity' of urban areas to accommodate more homes and the environmental implications of given levels of housing provision. In short, PPG 3 promotes the preservationist cause in its insistence that regional planning authorities should wherever possible prioritise the re-use of previously developed land and, more generally, concentrate residential development in urban areas so as to protect the countryside.

As we saw in Chapter 4, this policy is already beginning to have a significant impact in some regions. In the review of regional planning in the South East, a regional capability study was undertaken which showed that the region could only meet its housing requirements if it breached environmental constraints. This capability study thus gave rise to a scenario in which conventions of development and environment came into conflict with one another. Subsequent government attempts to finesse the policy (e.g. by choosing housing numbers midway between the region's total and the national forecast and claiming that the 'plan, monitor and manage' system outlined in PPG 3 means more houses can be provided on less land) have not diminished the central issue. Once development forecasts and strategies are immersed in complex regional contexts they inevitably lose their immutability and become subject to a host of competing considerations. At present, preservationist considerations seem to predominate in both national planning frameworks and the approaches being adopted at the regional tier of planning (Murdoch and Abram, 2002).

It thus appears as though the strengthening of regional planning will cut across the regional economic approaches of the RDAs, particularly in areas such as the South East where, despite the high levels of economic growth, the countryside is increasingly encompassed within a protectionist policy regime. Regional planning bodies are being charged with delivering sustainable regional plans that somehow harmonise the protection of regional environments with the facilitation of economic development. However, the emphasis emerging from PPG 3 and other such policy documents indicates that protection (rather than harmonisation) is to the fore so that development will need to work around valued natural assets. This holds particular implications for the countryside, which is often perceived (especially in places where environmental networks are active) as an environmental asset in need of protection. The economic development priorities of the RDAs may therefore be forced to accommodate preservationist conventions enshrined within regional plans.

Regional policy and the differentiated countryside

In Chapter 3 we proposed that regionalisation in the countryside is emerging from the interactions between three differing types of networks: *political* networks, which are stimulating the construction of a multi-level goverance system for rural areas; *economic* networks, which are encouraging the build-up of productive capacities and associations at the regional level; and *social* networks, which are coalescing into regional social formations as a result of social changes, such as counterurbanisation, and the associated political activities of social actors as expressed, for instance, in the environmental movement. These differing networks cover varied spatial scales and operate in a variety of ways. However, what they have in common is that all are making claims on rural space. Moreover, they are making these claims in sub-national regional contexts. The differential interactions between the networks in sub-national spaces gives rise, we have argued, to the differentiated countryside.

Now the regionalisation process has begun to affect policy. As we have outlined above, a previously national policy regime has begun to disaggregate into a set of regionalised approaches in which rural development and planning policies are formulated at the regional level. While national policy still plays a 'framing' role, a growing amount of rural policy is formulated and implemented in a regional structure of some kind. Moreover, not only is the region now able to formulate its own policies, but it is expected to do so by including as many regional stakeholders as possible in the review process. Thus, a form of 'inclusive' regional policy-making is starting to emerge and again this marks a shift away from the post-war approach, which involved only a limited number of (mainly agricultural) interest groups as participants.

Given this emphasis on partnerships and stakeholder involvement, we might assume that the development trajectories of the regions included in our case studies will be increasingly shaped by the networks that dominate rural land development processes. Thus, in the South East region we should expect that environmental networks will be to the fore in regional planning processes in order to ensure that the interrelated goals of urban containment and rural preservationism are placed at the centre of planning policy (Chapter 4 gives some clear indications that environmental networks are beginning to play this regional role). Moreover, evidence reported in Chapter 3 (drawing upon work conducted by Keeble and Nachum, 2002) indicates that many key economic interests based in rural areas will support countryside protection strategies, largely because such strategies protect the locational qualities that drew them into the countryside in the first instance. The main opponents of preservationism and containment will be developers, notably the house builders, who wish to gain access to land in a regional context where demand for housing and other land uses is high. The South East region therefore displays an uneasy balance between strong protection of the countryside and dynamic levels of urban growth. The balance will be codified in regional planning policies.

In the South West we can expect that regional policy processes will exhibit the same tensions between development for local needs and protection of the environment that we encountered at the county and district levels. The regional planning body will find itself caught between demands from the one side, to build new houses for both the local and the incoming population (so as to keep house prices within reach of the traditional rural population) and demands from the other, to protect the rural environment by keeping levels of new house building to a minimum. The RDA will similarly be beset by demands to provide employment and development opportunities for the traditional rural population, while at the same time being encouraged to protect rural environments and communities. Yet, although it appears as though the regional tier of governance simply provides another venue for the processes of contestation that are currently unfolding at the local level, there is clearly some scope to overcome the rather fixed positions associated with the development and environment networks. The emphasis on partnerships and joint working may facilitate some degree of coalition building in the South West so that a more nuanced policy emerges. This policy is likely to combine a recognition of both development and environment needs for the rural areas of the region and may lead to a mutual accommodation between agrarian and environmental networks so that greater attention is given to the distinctive 'environmental economy' of the South West.

In the North East the new regional policy fora are likely to be dominated by economic and social development concerns. Among economic planners, regional policy-makers and the region's political and industrial leaders there is a broad consensus on the overriding need to boost the region's economic growth and competitiveness. Sustainable development is seen in terms of 'achieving high and stable levels of economic growth that bring prosperity to everyone while protecting and improving the environment' (One North East, 2002: 11). Against the considerable concentrations of industrial decline and urban deprivation, the rural landscapes of the North East are seen to be of potential economic benefit as the region increasingly comes to recognise that it 'must capitalise on environmental, biodiversity and heritage assets to deliver sustainable economic growth' (ibid.: 59). As our landowner case studies show, there is much support for this approach among the rural elite, alongside a concern to ensure that development processes remain in keeping with social and environmental characteristics. Despite the fact that the new regional planning system apparently allows the environmental voice to be heard more forcefully in the region (so that organisations such as the CPRE invest resources in establishing regional structures – see Lowe *et al.*, 2001), it is likely that preservationist networks will continue to be perceived as 'external' actors with no ready constituency in the North East region.

Thus, the new regional policy fora would seem to provide an opportunity for dominant networks to consolidate their hold on rural policy-making. However, the new policy-making arrangements may also hold some scope for coalition building between networks and may allow previously marginal networks to gain some influence. New political alignments may emerge if elected regional assemblies come into being. In *Your region, your choice* (DTLR, 2002) the Government provides the opportunity for the English regions to comprise elected regional assemblies. These assemblies would take responsibility for economic development and planning, as well as transport, culture and housing. In the government's view, the assemblies would 'better reflect the needs of the region' by 'ensuring that relevant stakeholders are engaged in developing and delivering [the regional] strategies' (ibid.: 3). While it is unclear how many of these elected assemblies will ultimately come into being (the government suggests they will appear wherever there is a regional demand for elected bodies), they clearly hold some scope for dislodging dominant regional networks.

A key issue in assessing the likely trajectories of change in our case study areas no matter what the governmental arrangements is the means by which competing priorities are to be negotiated within and between the new regional bodies and agencies. As yet the government has not made clear how any conflicting aims and objectives are to be reconciled. In particular, government has been unwilling to indicate whether, in the wake of any conflict, the economic strategies of the RDAs should take priority or whether the environmental planning approaches of the regional planning bodies should prevail. It has simply said that the various agencies will be drawn together through 'mutual interest' which will entail them working 'constructively together' (DETR, 1999: 23). The government thus seems to be hoping that conflicts will not emerge due to a process of 'iteration' in regional policy making (see DETR, 1999: 22).

Not surprisingly, this unwillingness to specify more concrete arrangements has led many participants to worry that *either* the environmental *or* the economic agenda will prevail. The House Builders Federation (1998) believes that discussion of sustainable development strategies at the regional level will mean further rounds of preservationist planning, thereby further limiting development options. The CPRE worries that regional planning policy is simply not strong enough to impose a meaningful environmental policy. It argues:

> Clear objectives for maintaining the region's important environmental assets – such as landscape quality, countryside character, rural tranquility, and natural resources – have rarely been established in regional planning guidance and translated into effective policies which prevent their being gradually lost through the culmulative effects of numerous 'trade offs' whenever there is a conflict with other objectives, such as providing attractive greenfield sites for incoming business developments. This is a fundamental shortcoming. A primary purpose of the land use planning system is to manage the impact of development and to meet social, economic and environmental objectives . . . but this will not be possible unless planning policies act as a *real brake* on

> physical expansion into the countryside. . . . In the absence of clear objectives and policies to maintain a clearly defined *environmental 'bottom line'*, environmental assets will be vulnerable to incremental but unlimited loss.
>
> (CPRE, 2001: 9, emphasis added)

The way the differing agendas of developmentalism and environmentalism are evaluated both within the regions and by national government holds implications for patterns of differentiation in the countryside. If the RDAs take priority over the regional planning bodies then the policy of preservationism that has been so assiduously cultivated in local rural areas in the South East and parts of the South West would be threatened and the 'environmental bottom line' that many preservationist actors wish to see put in place would be pushed even further back. The countryside could thus be quickly transformed despite the undoubted strength of the opposition that would arise from counterurbaniser groups. In contrast, in the North East, where the RDA acknowledges that 'in the short term there may be tensions between achieving economic, social and environmental goals at the same time' (One North East, 2002: 75), the pressures on the rural environment remain much less intense despite the prevailing economic boosterism because, in the region overall, population is declining, economic growth rates are sluggish and development priorities are focused on urban regeneration.

If, on the other hand, the regional plans were to take priority over the RDAs then we could similarly see the imposition of rather restrictive planning policies in the South East and the South West. While the impact of this in the South East would be even more intensive urban growth, in the South West it may threaten efforts to diversify the economies of the more remote rural areas. In the North East, meanwhile, a strengthening of regional planning could enhance the leverage of regional environmental actors in a region governed in the main by economic conventions, but this may do little to alter the main trajectory of rural development given that it is most strongly influenced by the large private and institutional landowners.

Conclusion

The move away from a national agricultural policy loosens the grip of the national state on the panopoly of rural areas to be found in England. We now see the emergence of a broader rural policy concerned not only with agriculture but with the full range of economic activities in the countryside. Yet, despite a profusion of policy statements (such as the Rural White Papers), no strong regulatory framework has emerged to direct rural economic and social processes. We are thus forced to conclude that national policy-making in the countryside has been weakened by the declining significance of agriculture and agricultural policy.

In this context, we witness the emergence of regionalised policy initiatives. We have examined two in the preceding pages: the RDAs and the regional planning fora. The RDAs are now to act as the lead agencies in the sphere of rural economic development. Their main objective is to help in the regeneration of rural areas by building up the local economic capacities of such areas, in part by orchestrating local and regional networks. This is a new approach to rural economic management and it holds the potential to strengthen those rural economies that have been weakened by agricultural productivism and the concentration of dynamic economic sectors in towns and cities. By 'thinking regionally' the RDAs may be able to create economic synergies across rural and urban areas.

However, in order to play this role in ways that genuinely benefit the countryside, the RDAs will need to be attuned to the particular requirements of small rural businesses and the nature of the networks being created in rural areas. In particular, they will need to consider the relationship between rural economic activity and the state of the rural environment (as we saw in Chapter 3 many businesses are drawn to rural locations because of environmental features). In other words, RDAs will need to combine rural economic growth with protection and enhancement of rural environmental assets, a combination that is normally termed 'sustainable development'.

A 'sustainable' approach may be assisted by the activities of the regional planning bodies. These bodies are also being asked to 'think regionally' and to assess the significance of rural areas in a regional context. There is potential here for regional planning to address the economic, social and environmental needs of rural areas simultaneously. However, as we noted in Chapter 4, there is a continuing concern that regional planning may be steered towards preservationism. While a preservationist approach will ensure the protection of valued natural assets, it may do so at the expense of economic growth. It may also open up a divide between the RDAs and the regional planning bodies to the detriment of rural governance processes, not least because the central state may be forced to step in where conflicts arise.

If the new system is to work successfully each of the regional bodies must be given the freedom to develop rural policies that are tailored to the requirements of increasingly differentiated regional spaces. Moreover, these policies must reflect the various economic, political and social constituencies to be found in the respective regions (not just those constituencies that can 'shout the loudest'). As our case study chapters show, dominant networks in the countryside will attempt to dominate regional policy arenas. If the new regional bodies are to act on behalf of all those residing in their respective rural regions then they must develop regional governance mechanisms that are inclusive of a broad range of interests. If they can facilitate the emergence of such mechanisms then they may be able to initiate more innovative patterns of regional development than those suggested by the opposing narratives of preservationism and developmentalism.

In sum, policies for the differentiated countryside need to be underpinned by robust regional structures and supported by a benign central state. As Little (2002: 25) puts it, the creation of associational modes of governance in the regions and elsewhere 'requires impetus from the centre as well as from below. It is within the remit of the state to encourage the decentralisation of power and to ensure that communities have the capacity to take on greater responsibility'. We have seen in this chapter that the state is currently willing to bring new modes of regional governance into being. But this is only the first step for it is also evident that both central and regional agencies must work to ensure that local networks of all kinds are able to involve themselves in these regional fora. These agencies need to encourage a range of differing groups, networks and associations to engage in various forms of democratic activity. And in order to undertake this approach effectively, the regional institutions will need to be further democratised so that conflicts between networks (e.g. preservationists and developers) can be resolved to the benefit of the many and not just the few residing in rural regions. The effective governance of the differentiated countryside seemingly requires, therefore, that regional institutions are accountable to the full range of regional 'stakeholders' so that regional policies reflect a widely agreed framework of conventions and aspirations for future development trajectories.

8 The dynamics of differentiation

A decade has now passed since we began our work under the ESRC's Countryside Change Programme. The overall aim of the research has been to review the main economic, political and social trends taking place in the countryside and to consider the way these are reflected in processes of land development. Through a series of general theoretical statements and the application of these to a number of case studies we have sought to uncover the main dynamics of change in the countryside. The present volume represents the culmination of this research and provides an overview of the various themes pursued under the Programme.

In this final book we have focused upon the varied nature of the development processes unfolding in the countryside and we have set these processes within separate regions. In the preceding chapters we have described an undulating geography of networks and conventions and we have investigated ways in which these social forms are currently orchestrating a 'regionalisation' of rurality so that differing rural areas become increasingly differentiated from one another. We have seen economic, political and social networks interacting in varied ways to shape patterns of regional development. And we have described a changing policy structure that increasingly reflects this pattern of geographical differentiation.

We have therefore attempted to take further the initial characterisation of the 'differentiated countryside' presented in *Constructing the countryside*. Adopting the 'ideal types' as a starting point, we have sought to investigate the uneven development of rural areas and to identify the ways in which discrete geographical spaces emerge from processes of co-operation and competition within and between economic, political and social networks. This last book, then, can be seen as a continuation of the research agenda opened up at the beginning of the Programme and it develops many of the themes that were elaborated in *Constructing the countryside*, *Reconstituting rurality* and *Moralising the environment*.

In reflecting upon the research undertaken under the 'Countryside Change' umbrella we can say, first, that many of the trends identified in earlier books, such as the move away from agricultural productivism, the growth of the consumption countryside, the urban–rural shift of both population and industry, and the growing intensity of the contest between 'pastoralism' and 'modernism' (in the guise of preservationism versus development), have become even more significant in recent years. Culmulatively, these trends contribute to a shift away from a *national* rural space to a *regionalised* set of processes and outcomes. The resulting patterns of regionalisation appear to justify our earlier identification of the ideal types – the 'preserved', 'contested', 'paternalistic' and 'clientelistic' countrysides – and the analysis of these in the present book.

Second, we have attempted here to analyse the ideal types in ways that largely derive from the theoretical postulates presented in earlier work. In the first monograph, *Constructing the countryside*, we identified five main processes which we argued should guide theoretical and empirical investigation of the countryside. These were: *regulation*, *production–consumption relations*, *commoditisation*, *representation* and *property rights*. *Regulation* is a theme that has remained at the forefront of our concerns and it runs through the present book. We have outlined how national agricultural policy has gradually begun to give way to a broader rural policy in which agriculture is progressively brought within a spatial (rather than sectoral) frame of reference. We have also suggested that recent changes in the structure of government (notably the establishment of the RDAs and the strengthening of the regional planning bodies) indicate that rural policy is increasingly likely to be formulated and implemented at the regional scale.

Patterns of differentiation also reflect a changing balance between *production* and *consumption*. Most obviously, the use of rural land for production purposes is affected by the move away from a national agricultural regime. Agricultural land is increasingly encompassed within a 'market-led' approach which leads to the emergence of 'hotspots' of agricultural productivism as well as areas of more extensive production. The urban–rural shift of manufacturing and services creates its own pressures for land development. In certain (accessible) rural areas – especially where new business agglomerations are evident – the demand for commercial land tends to be high, and places considerable pressure on the planning system (the case of South-East England illustrates this trend). In other, more remote rural areas there may be sufficient commercial and industrial land in relation to limited demand. Here the constraints on its use will lie more in the depressed nature of the surrounding regional economic context than in regulatory restrictions (the case of North-East England comes to mind here). Other areas lie between the two extremes with certain regions witnessing industrial pressure and industrial decline simultaneously. Land development processes will unfold accordingly, giving rise to a geography of uneven economic development within a single region (arguably the case in the South West).

Consumption relations also vary across the countryside. In accessible locations (such as Buckinghamshire and parts of Devon) new residents assert uses of rural land which are strongly determined by their residential preferences (i.e. the desire to live in a 'green' and 'quiet' rural location). They therefore resist overly economistic and developmental processes of change. In such locations the conflicts around land development take the form of a contest between development networks and preservationist networks and this contest runs through processes of regulation, notably planning. In more remote rural areas (yet other parts of Devon and much of Northumberland), the consumption countryside associated with the residential middle class is less evident. Development and economic networks, including tourism, therefore enjoy more freedom to pursue their own developmental agendas (which may be strongly tailored to the regional context, as is the case in Northumberland).

Commoditisation refers to the advance of market relationships into rural activities. It is most evident in economic processes where varied market relations are asserted according to the prevailing production–consumption linkages. Clearly the changing nature of the rural economy implies changing forms of commodity production and exchange (broadly a shift from primary to manufacturing and service activities) and these will be reflected in changing land uses. Commoditisation is also associated with counterurbanisation as the movement of middle-class households into rural areas means that the rural communities and environments are increasingly assessed within private housing markets (a process that is perhaps most advanced in Buckinghamshire and parts of Devon). A second mode of commoditisation is associated with the transformation of rural resources into touristic environments (this affects Devon and Northumberland more than Buckinghamshire, given that a fully counterurbanised rural environment is often of little interest to tourists).

A fourth concept presented in *Constructing the countryside* is *representation* and this refers to the ability of actors and groups to be heard within political processes. Again, this notion has been implicated in the research findings presented above. For instance, in Chapter 4 we illustrated how planning processes in Buckinghamshire and the South-East region allow for the expression of both developmental and preservationist views and values. Such expressions are associated with groups of actors that work to represent particular constituencies and particular environments. Likewise, in Devon we found a range of groups claiming to 'represent' the rural region. Indeed, a conflict over the legitimacy of differing representations was to the fore in this location, with both agrarian and preservationist networks jostling for supremacy. In Northumberland, much less conflict was evident. Thus, the large landowners and other developmental actors were effectively able to 'speak for' this traditional rural area.

Lastly, *property rights*. Throughout our work we have maintained an analytical focus upon rural property rights as a central organising framework. Initially, in *Constructing the Countryside* we speculated that property rights are an important 'window' through which the balance of economic, political and social forces in the countryside might be perceived. We have again proposed here that a broad sociological interpretation of property rights permits the assessment of power relations in the contemporary countryside. Thus, in Buckinghamshire we find the middle class is able to dominate social and political institutions and this domination effectively extends to the ownership and uses of property. In Devon, the balance of property rights reflects the balance of political forces between agrarian and preservationist interests, with some land assets being subject to a developmental rationale while others are protected. In Northumberland the power associated with the landowners' property rights are clear; here, property rights extend beyond the estates to other forms of economic and social control across the region.

While we have developed our analysis in line with these earlier theoretical notions, we have embellished our approach by adding *networks* and *conventions* into the theoretical 'mix'. We introduced the notion of the network in order to bind 'external' and structural influences on land development to 'internal' and local actions in the development process. Thus, we proposed that the network concept might function to integrate the 'general' and the 'particular' within a coherent analytical framework. Moreover, we also suggested that the network approach might allow us to consider how economic, political and social processes interact with one another and the way the combined operations of these processes leads to a regionalisation of the countryside. Thus, within the differing regions studied in the preceding chapters, we have attempted to identify the dominant networks and to consider how actors become aligned within them.

In examining the alignment of actors we have paid particular attention to conventions, that is, to the agreements that allow actors to match their individual goals with the goals of the networks. We have identified differing convention types and have suggested how these might be combined in the various networks that act to configure regions in the differentiated countryside. In some networks (notably those associated with pastoralism or preservationism) local, civic and ecological conventions are to the fore and these are embedded in rural communities and the rural environment (Buckinghamshire provides the clearest example here). Other networks highlight economic and developmental conventions and push to ensure that land is made available for the requisite activities (all three regions hold illustrations of this process). In yet other networks we find an accommodation between developmental and local ecological conventions so that the two approaches are brought into some kind of alignment (the landed estates in Northumberland illustrate how this can be achieved). Thus, we find conventions becoming eshrined within socio-material formations in the countryside as networks seek to harness economic, political and social trends to particular regional development trajectories. These formations therefore define the various landscapes of the countryside.

The study of networks of conventions in the countryside also allows us to suggest that the regionalised ruralities under scrutiny here represent different compromises between forms of mobility in contemporary English society (transportation, residential migration, tourism, flexible industrial strategies, etc.) and the socio-material 'endowments' of given rural spaces (workforces, communities, buildings, ecological features, etc.). These 'compromises' might be assessed using Urry's conceptualisation of place-making, in which

> places can be loosely understood . . . as multi-plex, as sets of spaces where ranges of relational networks and flows coalesce, interconnect and fragment. Any such place can be viewed as the particular nexus between, on the one hand, propinquity characterised by intensely thick co-present interaction, and on the other hand, fast flowing webs and networks stretched corporeally, virtually and imaginatively across distances. These propinquities and exten-

> sive networks come together to enable performances in, and of, particular places.
>
> (2000: 140)

In the rural context these 'performances' have highly differentiated social, economic and political outcomes and thus vary from place to place. In each of the case study regions we have shown how conventions and networks become 'set in space'. Over time such 'settings' give rise to distinctive socio-spatial formations, formations that are embedded in the material environments of given rural regions.

We have also suggested that rural regions emerge from *reflexive* practices as convention hierarchies are constructed and imposed within mobile networks. Again, reflexivity is more evident in some places than others. It also takes a variety of forms. In Buckinghamshire we find the archetypal middle-class forms of reflexive practice associated with particular constructions of the community and the environment. These forms are dominant in this strongly counterurbanised area. In Devon, however, middle-class reflexivity mixes with tradition, notably agrarianism. In many respects the reflexive networks of the middle-class incomers and the agrarian networks of long-established residents are separate and distinct from one another. Nevertheless, some amount of coalition building between the two is beginning to take place. Thus, we may begin to witness the emergence of a reflexive mode of 'agrarianism', one that somehow aligns the aesthetic concerns of the new residents with the more developmental perspectives of traditional groups. Lastly, in Northumberland a form of reflexive stewardship can be discerned as new landlords consciously adopt the mantle of paternalism. Arguably, in this region reflexivity turns full circle: in recent decades the middle class has come to adopt earlier aristocratic perspectives in order to develop its own form of aesthetic reflexivity; now the landlords and aristocrats adopt forms of middle-class aesthetic reflexivity in the management of their estates (for example, whereas the castles and great houses of the traditional landowners dominated the scene, the homes of the new owners tend to be secluded, reflecting norms of privacy).

This interaction between reflexivity and tradition clearly holds implications for the relationship between pastoralism and modernism outlined in the introduction. As we indicated above, the perspectives of pastoralism and modernism have defined the character of the English countryside for a century or more. Yet, the balance between the two is constantly changing. For instance, the modernising impulse that has run through practices and policies of agricultural development in the second half of the twentieth century has given way to a much more pragmatic (market-oriented) approach. As we saw in the previous chapter, economic modernisation now increasingly applies to *rural* and *regional*, rather than *agricultural*, economies.

While the process of modernisation has perhaps been modified, pastoralism has remained a powerful force. Today's differentiated countryside is an expression of the seemingly relentless demand on the part of many households to live in an aesthetically-pleasing rural environment (in order to escape from the perceived 'anomie' of the dense, urban, metropolitan environment). A major driving force in contemporary processes of change thus concerns the quest on the part of many individuals and families for *the rural*, both as an experience (in the form of countryside tourism) and as a 'refuge' from modernity (in the form of a house in the country). In this sense, the cultural signifance of a traditional 'urban–rural dichotomy' is alive and well. However, this broad spatial distinction is also being regionalised. Thus, the differential and multiple imposition of diverse values, codes and conventions in differing social and economic spaces is taking place within the new regional contexts.

In Buckinghamshire the division between the urban and the rural – which is seemingly transgressed by the counterurbanisation process – is reinforced by the politics of preservationism. Urban areas therefore grow while rural areas are protected. In Devon, counterurbanisation promotes a preservationist politics that also seeks to maintain strict divisions between urban and rural spaces. However, it is here confronted by representatives of an agrarian social formation that appear to regard the distinctively rural characteristics of the area as only viable if some amount of development is allowed to take place. Agrarian networks, therefore, appear more relaxed about maintaining any sharp distinction between urban and rural locations. Lastly, in Northumberland the landowners and other influential rural activists see the rural occupying a particular cultural space within an urban and industrial region – it is a reservoir of tradition and retains many distinctive social and environmental characteristics. However, it must be allowed to develop in order to maintain a working population. Again, the boundary between urban and rural is a little more fluid than in the preserved countryside, although, ironically, the amount of development taking place is much less marked, meaning that existing rural characteristics are more easily maintained.

In the three areas, then, we see that the boundary between metropolitan areas and the countryside being re-defined in keeping with the emerging regional contexts and the associated hierarchies of conventions. And as regionalisation advances, and as the distinctions between regions increase, we should expect to see new urban-rural relations coming into being, perhaps more tailored to regional circumstances than has been the case under the post-war productivist regime.

These observations therefore lead us to ask, in conclusion, what the future trajectories of development of our three regions might be. In Buckinghamshire the struggle to preserve rural assets in a dynamic regional context will undoubtedly continue. Counterurbanisation remains a dominant trend in the region and this will reinforce the middle-class character of the countryside. Yet, counterurbanisation also creates opportunities for developers (notably in the housing field) and they seek to pursue these through the planning system. While we have suggested that stronger regional planning may promote stronger forms of preservationism there is still the possibility that central government direction (and the continuing concern to ensure high levels of economic growth, especially within the strategically significant South East region) may work to overturn preservationist tendencies. The policy pendulum seems at present to be swinging towards preservationism; yet, it may equally swing back towards a stronger form of economic modernisation (see for instance DTLR, 2001).

Devon holds some key characteristics in common with Buckinghamshire, notably the high levels of counterurbanisation. Again, these look set to continue so that more and more of Devon's countryside takes on a middle-class character. However, the middle class here confronts a rather traditional agrarian social formation, one that is rooted in Devon's distinctive economic and social history. The interaction between these two social formations will define the future of rural Devon. Two scenarios suggest themselves: on the one hand, the two sets of networks will continue to contest development patterns in ways that promote continuing conflict in political and social arenas; on the other hand, they will evolve a mode of mutual accommodation in order to establish a regulatory approach that is in keeping with Devon's rather distinctive character. The emergence of stronger regional policy fora may facilitate the realisation of either of these scenarios: the regional bodies may be used to simply carry on local patterns of contestation or they may facilitate a genuine dialogue between the networks. Out of this dialogue may come a new form of politics that moves beyond agrarianism and perservationism.

In Northumberland, counterurbanisation is likely to be only a marginal influence for the forseeable future. Therefore, barring any major national government assault upon the landowning classes (most unlikely at the present time), the estates are likely to continue to shape the overall development trajectory of much of the rural region. They will not only continue to play a significant role in influencing land development agendas, but also, by dint of their cultural as well as physical stake in the landbase, will articulate a renewed form of 'stewardship', one that combines aesthetic, ecological, localistic and economic conventions. Thus, rural Northumberland is likely to maintain its own distinctive characteristics and these will be embedded not only in the large landed estates but in the economic, social and environmental characteristics of the wider countryside.

To sum up, we can say that the development trajectories that we have identified in the differentiated countryside are, to a large extent, *path-dependent*. They result from the interactions between economic, political and social networks. In these interactions local actors align their aims and objectives with the broader networks so as to promote particular sets of conventions. These conventions help to determine future patterns of economic development, the degree of protection to be enjoyed by rural environments, the strength of rural communities and the character of the wider countryside. In other words, they demarcate the most significant features to be found on the landscapes of England's increasingly differentiated regions.

We can therefore conclude that the countryside is no longer (if it ever was) a single, stable and coherent entity: rather, it is multiple and diverse in character; it encompasses a host of differing economic institutions, political networks, social groupings and environmental features. These combine in differing ways in differing places. There is, then, not just one countryside but a multitude of country*sides*, each changing in line with a specific set of time–space co-ordinates. While this finding may seem to undermine many taken-for-granted assumptions about this quintessential English territory (assumptions that derive, in the main, from the narratives of pastoralism and modernism), it has the virtue of representing rural areas in England (and elsewhere) as places where various forms of identity can be performed, where diverse trajectories of development can be negotiated and where multiple modes of evaluation can be expressed. These changing characteristics increasingly serve to define the differentiated countryside.

Notes

1 A differentiated countryside?

1 Despite the fact that, in reality, employment in the countryside has been relatively buoyant. Compared to urban and metropolitan areas, rural local authority districts experienced higher employment growth between 1980 and 2000. Some 590,000 net new jobs were created in rural areas between 1980 and 1990 and a further 600,000 between 1990 and 2000 (Public and Corporate Economic Consultants, 2000, para 6.4.2).
2 This is not to say that agriculture is insignificant in all locations. In some of the most agricultural localities, farming can be found to employ much larger proportions of the local workforce: 22 per cent in the district of Eden in Cumbria; 25 per cent in South Holland District in Lincolnshire; 31 per cent in Leominister; and 33 per cent in Radnor in Wales (MAFF, 1999)
3 However, as we will outline on pp. 23–8, we do draw upon an approach to comparative cultural *sociology* derived from conventions theory (see Lamont and Thevenot, 2000).
4 It should be noted that the typology was not intended to be exhaustive, nor was it claimed that any rural locality can simply be placed exclusively in one of the 'types' (a locality may display a mixture of characteristics, as our case studies in this book will show). Instead, the typology was intended for use as an analytical starting point, one that helps draw attention to divergence in the fortunes of rural areas. It indicates that new patterns of differentiation, based upon locally and regionally discrete economic and social formations, are emerging.
5 Although, just to confuse matters, Scotland, Wales and Northern Ireland are often seen as simultaneously countries *and* regions of the UK.
6 Keating believes decentralisation in European nation-states is also encouraged by the European Union: not only are territories now competing within a single European market but European Union funds for diminishing the disparities between territories are increasingly administered on a regional basis. This is leading to much talk about a 'Europe of the regions' (Balchin *et al*., 1999; Jonsson *et al*., 2000). As we shall see in Chapter 2, this process has impacted upon rural areas mainly through the Common Agricultural Policy and the Structural Funds.
7 In line with this approach, we use a flexible interpretation of the 'region' in our analysis of rural differentiation in the following chapters. Our case studies span the region/locality divide as they refer to English 'shire' counties within their regional contexts.
8 In the work of Castells and others we find general network theories that seek to account for global patterns of change but make no particular reference to the 'rural'. It must, therefore, be inferred that rural space is simply encompassed within the architecture of the global network and that it is likely to sit rather low in the hierarchy of network 'nodes'. For instance, the urban concerns that lie at the centre of Castells'

(1996) work lead us to assume that the countryside lies a long way from the organising centres of the networks and therefore that it is peripheral to network society. In the following chapters we aim to correct this view by showing how the rural is bound into the new regional networks.

9 Thevenot *et al.* (2000), following Boltanski and Thevenot (1991), claim that the identification of these convention types follows from empirical observation in a range of differing contexts, although it also seems to owe much to work of political philosophers such as Charles Taylor and Michael Walzer. In other words, there are multiple spheres of justice or evaluation but this multiplicity is not endless: it is circumscribed by the available modes of evaluation prevailing in any given social context. While in our own study we pay some heed to Thevenot *et al.*'s types we do not follow their typology rigorously although clearly we pay heed to the conventions they identify (for a varied list of convention types see Storper and Salais, 1997).

10 In this sense, conventions refer to a 'blackboxing' of network relations (as described by Latour, 1987); that is, various interactions and connections are simplified into convention modes that act to both describe and reflect these interactions and connections. Conventions and network relations are therefore always in some kind of (often creative) tension with one another.

2 Regionalising the rural

1 Although a Farmers' Union of Wales (FUW) was formed in the mid-1950s as an alternative representative organisation to the NFU for the small farmers of Wales it was excluded from the Annual Review for over twenty years. This exclusion was largely as a result of NFU lobbying of central government. By excluding its competitors from the Annual Review the NFU could continue to claim that it was the farmers' voice inside government (on the rise of the FUW see Murdoch, 1995a).

2 The effects of such policies became evident once government began to collect data on regional environments. For instance, DEFRA's (2002b) publication *Regional quality of life counts* shows that between 1970 and 1999 bird populations in England declined by between 11 and 21 per cent in the East of England, the South East, the South West and Yorkshire and Humberside, while they increased by between 27 and 50 per cent in the East and West Midlands, the North East, and the North West.

3 It has been argued, by for instance the Cabinet Office Performance and Innovation Unit (PIU), that 'over the longer term, as price support and compensation payments to farmers are reduced, so some of the savings can be re-directed to environmental and rural development schemes through expanded use of the RDR' (PIU, 1999: 92).

4 The Barlow Report refers to the Royal Commission on the Distribution of the Industrial Population which recommended a 'controlled' dispersal of industry, especially to underdeveloped regions. This emphasis on 'control' permitted both regional policy and planning to play a role in the location of industrial premises.

5 The introduction of planning powers meant that planning authorities (effectively the counties and county boroughs) were required to fulfil two tasks: to prepare development plans for land use in the local authority area ('forward planning') and to control land development in the context of the plan ('development-control regulation'). All proposals for development were to be submitted to the planning authority and were to be judged in accordance with the plan. Decisions were to be made by elected councillors working on advice from planning officials. According to Hall *et al.* (1973), this decision-making structure reinforced the urban–rural divide as the county councillors invariably acted in favour of rural protectionism.

6 In essence, these developments were a private-sector version of the public-sector growth centre approach that had prevailed during the 1950s and 1960s in the new and spillover towns.

7 The significance of this measure was heightened by the widely held belief that the

previous Secretary of State for the Environment had been planning to abolish structure plans thereby weakening the forward planning system.

8 Of course, commuting is not a particularly rural phenomenon. The majority of those who are economically active travel out of the ward in which they live in order to work, and this is in fact more generally so in urban than in rural wards. Rural commuting, though, is marked by certain tendencies: those living in rural areas who do commute tend to travel much further and are more likely to commute by car. In consequence, long distance commuting is a particular feature of social stratification in rural areas, distinguishing those who operate in urban and regional labour markets from those dependent on local employment. The group that is more likely than others in rural areas to commute a long distance to work is long-distant migrants to the area, which indicates a link between counterurbanisation and commuting.

9 The Government Offices for the Regions were introduced by John Major's Conservative government in 1994. They comprised a new structure of ten integrated offices. Regional civil servants in the Departments of Employment, Environment, Transport and Industry were made accountable to one Regional Director who was to be responsible for all staff and expenditure. A set of overall objectives was established for the Offices, including requirements to: achieve operational demands of departments and ministers; contribute local views to the formation of government policy; promote a coherent approach to competitiveness, sustainable economic development and regeneration; and develop partnerships between local interests to secure these objectives (see Mawson and Spencer, 1997). Initially, MAFF remained separate from the Government Office structure.

3 Theorising differentiation

1 The 'top-down' nature of planning policy is also evident in the use of national housing projecions to guide levels of house building so that individual local plans must include housing land allocation 'targets'. For more on this issue see Chapter 4 (also Murdoch and Abram, 2002; Vigar *et al.*, 2000)

2 Moreover, this class had few affiliations to traditional rural activities such as agriculture, thus an agricultural community of interest emerged in the wake of counterurbanisation (Newby, 1979).

3 It also – although this was neglected at the time – arose out of particular gender, ethnic and racial divisions and the way these divisions were expressed in rural areas (see Valentine, 2000).

4 Lash and Urry describe the process of re-evaluation in the following way:

> Pre-modern space was filled with the markers of place. It was filled with and only recognisable by social practices. These social practices were in effect place markers. It was space not dominated by the temporal element; that is, space not to move through, but to live in. . . . Modernity witnesses the emptying out of place markers and the greater development of abstract space. . . . Modern space is objective space, as subjectively significant symbols are emptied out. Reflexive modernity [or what might be more generally referred to as 'post'-modernity] though is accompanied by a re-subjectivisation of space, only in reflexive form.
>
> (1994: 55)

It is tempting to assume that this general shift only applies to urban areas, with their clearly identifiable modernist and post-modernist buildings and spaces. However, it is clearly possible to argue that rural areas have been subject to the same overall logic.

5 It is perhaps ironic, however, that in the view of some commentators the social transformation of the rural population markedly changes the nature of rural entities that are so valued. For instance, Bauman believes the traditional community cannot survive its reflexive appropriation by the new middle class. He says,

> Since 'community' means shared understanding of the 'natural or 'tacit' kind, it won't survive the moment in which understanding turns self-conscious, and so loud and vociferous. . . . Community can only be numb or dead. Once it starts to praise its unique valour, wax lyrical about its pristine beauty and stick on pearly fences wordy manifestoes calling its members to appreciate its wonders and telling all the others to admire them or shut up – one can be sure that the community is no more (or not yet, as the case may be). 'Spoken of' community (more exactly: a community speaking of itself) is a contradiction in terms.
>
> (2001: 11–12)

6 This, despite the fact that many counterurbanisers clearly depend on dynamic economic contexts in order to gain the wherewithal to live in the more exclusive rural locations (see Charlesworth and Cochrane, 1994).

4 The 'preserved countryside'

1 One mechanism for this is, of course, house prices: as fewer houses are built at the same time as demand for a rural home rises so prices inexorably rise pushing country dwellings beyond the means of all but the most affluent.
2 Unless otherwise indicated, the quotations provided in the following three chapters are taken from fieldwork interviews (see Acknowledgements for details on the respective projects).
3 In fact, the objective of rural protection, which was formulated right at the start of the review, was virtually adopted wholesale from the previous version of the Structure Plan, indicating that once a set of conventions has been entrenched in local planning it becomes hard to shift.
4 As part of the plan review process the County Council must call a public examination (the EiP) which is overseen by an independent inspector (chosen by government). In the EiP, all interested parties come together to debate the various provisions of the draft consultation plan document. In the light of the debate, the inspector will then propose his/her amendments which must either be incorporated or repudiated by the County Council.
5 In countering the developers' arguments, the County Council claimed that its own forecasts of future housing demand in the county were in accordance with national and regional totals. It thus claimed that local figures gained their legitimacy from the national and regional housing projections. The County claimed it had merely translated these higher figures into local demand projections. In short, it brought the whole planning-for-housing network to bear in order to repudiate the developers' arguments.
6 The nature of the debate at the EiP meant that 'non-planners' had some difficulty getting heard. Despite the fact that many of the other local participants had some amount of planning expertise (something which marks them off from most members of the 'public') they were unable to render their concerns into the technical language of housing numbers and allocations. Thus, they were marginalised during the main discussion of housing figures (for more detail on this point, see Murdoch and Abram, 2002).
7 These negotiations were also framed by a planning model, commissioned from private consultants to indicate the optimum distribution of new housing between urban and rural areas, with particular emphasis placed on transport considerations. This model was used to create a division of the housing figures between urban and rural areas.
8 The proposal to distribute development around the rural areas of the Vale thus confronted the Council with a new, fortified network of many rural villages waging a campaign against the Council's revised planning strategy of dispersed growth.
9 It has now been replaced by a 'regional assembly' made up of regional 'stakeholders' including local politicians, business representatives, interest group members and 'third sector' organisations.

10 In seeking to use localistic and environmental conventions (in the context of the regional capability study) to dislodge conventions of demand, SERPLAN had the backing of its constituent local authorities.
11 The projection of 4.4 million new households was produced by the Office of National Statistics during John Major's prime ministership and caused a huge political row to break out in planning circles. Environmental groups feared that this figure would run through the planning framework thereby facilitating something of a development 'free for all' in the housing sector. The Major Government responded to this political pressure by calling for a 'national debate' on the implications of the figures. This debate was still running when Tony Blair's Labour government came into office in 1997.
12 Because the CPRE is based on a county branch structure with a London headquarters it is able to lobby both locally and nationally simultaneously. Thus, the CPRE in London could argue for a change in national policy while the local county branches opposed the cascade of figures down the planning hierarchy. This structure also meant that the CPRE could orchestrate protests at the local county level (Lowe *et al.*, 2001).

5 The 'contested countryside'

1 In 1999 Cornwall received Objective 1 status, confirming it as one of the poorest regions in the EU, while most of Devon received Objective 2 status which is for areas that are undergoing rural or industrial re-structuring.
2 The county attracts by far the largest tourist expenditure from UK residents, with the total standing at about £1 billion a year, directly generating about 32,500 jobs (Devon County Council, 1999). Rural tourism is an important feature, with 44 per cent of the bedspaces being outside the resorts and urban centres (Devon County Council, 1998b, table 6). The sector is made up mainly of small-scale family businesses.
3 It should be noted that in-migration has had a marked impact on the age structure of the county which, in the words of Devon County Council (1998c: 71), is 'tending towards the more elderly'. More than a quarter of the county's population was aged 60 or over in 1996, significantly higher than the UK average of 18 per cent. In some districts, the dominance of older people is even more pronounced, particularly along the southern coastal zone where retirement is a key factor in maintaining local economies (Gripaios, 1994; Glyn-Jones, 1975). For example, in 1991 four out of ten households in East Devon District were occupied solely by pensioners (Devon County Council, 1998c). The majority of incomers to the county, though, are working-age people seeking to take up employment or business opportunities, although often with the eventuality of retirement in mind (Halliday and Coombes, 1995).
4 The Council did not appoint its first Chief Planning Officer until 1958, and even then it was at government insistence as a condition for approving the County's first Development Plan.
5 It also means that localistic concerns are refracted through a discourse of environmentalism (rather than the other way round). This can be contrasted with the developmental network which sees economic issues through a discourse of localism. As we show on pp. 105–8, the differential understandings of the 'local' lead to a fissure running through Devon politics.
6 The planners often appear to sit uncomfortably between the two sides: they (usually) recognise the validity of the professional arguments employed by CPRE but must also respond to the (local) sensitivities of the political members.

6 The 'paternalistic countryside'

1 In the North East as a whole, the proportion of the working population in professional, managerial and technical occupations is just 23.5 per cent, compared with an average for England of 30.6 per cent (Regional Trends, 1998: 48).

2 Analysis of the 1991 Census Special Migration Statistics by Champion *et al.* (1998) found a rate of net migration from Greater London of 7.7 per thousand in the preceding year, while net migration from Tyne and Wear was just 1.5 per thousand – the lowest rate of all metropolitan areas.
3 For instance, in the year 2000 Labour was the majority party (41 members) with Conservatives in second place (15 members) in Northumberland County Council.
4 A survey of almost 200 households (a 40 per cent sample) in the area found over 20 per cent of respondents in tourist-related employment, and widespread general support for the development of tourist facilities which had widened local economic opportunities.
5 It is interesting to note that the non-farming resources that are channelled into this paternalistic piece of rural Northumberland are derived largely from volume house building, an activity that has been instrumental in changing rural areas elsewhere.
6 Kielder village, for example, planned around the Duke of Northumberland's former hunting lodge, was originally envisaged to house 2,000 families including unemployed people re-located from the industrial conurbations. However, the introduction of the chainsaw meant that even with the expansion of the Forest, employment never reached these levels, and the village stands as a monument to a particular form of state-directed rural industrial development. The village now suffers acute problems associated with peripherality, depopulation and loss of services.
7 For example, one landowner confided that when the Ministry of Defence (MoD) had been considering a ban on fox hunting over its land he had written to point out that he allowed the army to do exercises over his land and therefore it would be inadvisable for the MoD to exclude the hunt.
8 It is indeed arguable that the weakness of the CPRE's local base helps to give it clarity in its planning representations as there is not the need to reconcile the outlook of a lot of active members.
9 The same official argued that of the recent 'big cases', the CPRE took 'a keen interest' in the Great North Park proposal for housing development in the Newcastle Green Belt. However, the CPRE's involvement in the Great North Park case was exceptional. It has been much less active in what is perhaps the most significant planning issue to have arisen recently in Northumberland: the MoD's proposals for the Otterburn Training Area in the Northumberland National Park. In this case, the CPRE did not take the lead, but lent its support to the campaign of opposition by the Council for National Parks.
10 Drawing on data from county councils, Paxman comments:

> Thirty years ago, the aristocracy were mainstays of local government. In 1960 you could find four dukes, one marquess, nine earls, four viscounts, five viscountesses, twenty-seven barons, thirty-four baronets, fifty-two knights and fifteen titled wives sitting on English county councils. Today they have all but vanished.
>
> (1991: 45)

References

Abram, S., Murdoch, J. and Marsden, T. (1996). The social construction of 'Middle England': the politics of participation in forward planning. *Journal of Rural Studies* 12, 353–364.

Allanson, P. (1992). Farm size structures in England and Wales, 1939–1989. *Journal of Agricultural Economics* 43, 137–148.

Allen, G. (1959). The NFU as a pressure group. *Contemporary Review* May/June, 265–332.

Allen, J., Massey, D. and Cochrane, A. (1998). *Rethinking the region*. Routledge, London.

Allmendinger, P. and Tewdwr-Jones, M. (1997). Post-Thatcherite urban planning: a Major change? *International Journal of Urban and Regional Research* 21, 100–116.

Allmendinger, P. and Tewdwr-Jones, M. (2000). New Labour, new planning? The trajectory of planning in Blair's Britain. *Urban Studies* 37, 1379–1402.

Amin, A. (1999). An institutionalist perspective on regional economic development. *International Journal of Urban and Regional Research* 23, 365–378.

Amin, A. and Thrift, N. (1994). Living in the global. In A. Amin and N. Thrift (eds). *Globalisation, institutions and regional development*. Oxford University Press, Oxford.

Asby, J. and Midmore, P. (1996). Human capacity building in rural areas: the importance of community development. In Midmore, P. and Hughes, G. (eds). *Rural Wales: an economic and social perspective*. Welsh Institute for Rural Studies, Aberystwyth.

Atkins, D., Champion, T., Coombes, M., Dorling, D. and Woodward, R. (1996). *Urban trends in England*. Department of the Environment, London.

Aylesbury Vale District Council (1991). *Rural areas local plan*. Aylesbury Vale District Council, Aylesbury.

Aylesbury Vale District Council (1998). *District local plan: consultation draft*. Aylesbury Vale District Council, Aylesbury.

Baker, M. (1998). Planning for the English regions: a review of the Secretary of State's Regional Planning Guidance. *Planning Practice and Research* 13, 153–169.

Balchin, P., Sykura, L. and Bull, G. (1999). *Regional policy and planning in Europe*. Routledge, London.

Barclay, C. (1999). *British farming and the reform of the Common Agricultural Policy*. House of Commons, London.

Barlow, J. and Savage, M. (1986). Conflict and cleavage in a Tory heartland. *Capital and Class* 31, 32–56.

Bauman, Z. (2001). *Community: seeking safety in an insecure world*. Polity, Cambridge.

Bell, C. and Newby, H. (1972). *Community studies*. Allen and Unwin, London.

Bell, M. M. (1994) *Childerley*. University of Chicago Press, London.

Bennett, K., Caroll, T., Lowe, P. and Phillipson, J. (2002). *Coping with the crisis in Cumbria: the consequences of Foot and Mouth disease*. Centre for Rural Economy, Department of Agricultural Economics and Food Marketing, University of Newcastle.

Blokland, T. and Savage, M. (2001). Networks, class and place. *International Journal of Urban and Regional Research* 4, 221–227.

Boltanski, L. and Thevenot, L. (1991). *De la justification: les economies de le grandeur*. Gallimard, Paris.

Bolton, N. and Chalkley, B. (1990). The population turnaround: a case study of North Devon. *Journal of Rural Studies* 6, 29–43.

Bowers, J. (1985). British agricultural policy since the Second World War. *Agricultural History Review*. 33, 66–77.

Bowers, J. and Cheshire, P. (1983). *Agriculture, the countryside and land use*. Methuen, London.

Boyle, P. and Halfacree, K. (eds) (1998). *Migration into rural areas*. John Wiley, London.

Bradley, T. and Lowe, P. (1984). *Locality and rurality*. Geo Books, Norwich.

Brassley, P. (2000). Output and technical change in twentieth-century British agriculture. *Agricultural History Review* 48, 60–84.

Breslau, D. (2000). Sociology after humanism: a lesson from contemporary science studies. *Sociological Theory* 18, 289–307.

Brouwer, F. and Lowe, P. (2000). *CAP regimes and the European countryside*. CAB International, Wallingford.

Buckinghamshire County Council (1994). *County structure plan: consultation draft*. Buckinghamshire County Council, Aylesbury.

Butt, R. (1999). The changing employment geography of rural areas. In M. Breheny, (ed.). *The people: where will they work?* TCPA, London.

Castells, M. (1996). *The rise of the network society*. Blackwell, Oxford.

Champion, T. (ed.) (1989). *Counterurbanisation: the changing pace and nature of population deconcentration*. Edward Arnold, Sevenoaks.

Champion, T. (1994). Population change and migration in Britain since 1981; evidence for continuing deconcentration. *Environment and Planning A* 26, 1501–1520.

Champion, T. (1996). *Migration between metropolitan and non-metropolitan areas: a report for the ESRC*. Economic and Social Research Council, Swindon.

Champion, T., Atkins, D., Coombes, M. and Fotheringham, S. (1998). *Urban exodus*. Council for the Protection of Rural England (CPRE), London.

Champion, T. and Townsend, A. (1990). *Contemporary Britain: a geographical perspective*. Edward Arnold, Sevenoaks.

Charlesworth, J. and Cochrane, A. (1994). Tales of the suburbs: the local politics of growth in the South East of England. *Urban Studies* 31, 1723–1738.

Clapson, M. (2000). The suburban aspiration in England since 1919. *Contemporary British History*. 14, 151–174.

Cloke, P. (1983) *An introduction to rural settlment planning*. Methuen, London.

Cloke, P. (1985). Counterubanisation: a rural perspective. *Geography* 12, 13–23.

Cloke, P. (1989). Rural geography and political economy. In R. Peet and N. Thrift (eds). *New models in geography, Vol 1*. Unwin Hyman, London.

Cloke, P. (1993) Review of Marsden, T. *et al*. 'Constructing the Countryside'. *Times Higher Educational Supplement* 45, 22.

Cloke, P. (1997). Country backwater or virtual village? Rural studies and the cultural turn. *Journal of Rural Studies* 13, 367–375.

Cloke, P. and Goodwin, M. (1992). Conceptualising countryside change: from post-

Fordism to rural structured coherence. *Transactions of the Institute of British Geographers* NS 17, 321–336.

Cloke, P. and Little, J. (1990). *The rural state*. Clarendon, Oxford.

Cloke, P. and Little, J. (eds) (1997). *Contested countryside cultures*. Routledge, London.

Cloke, P. and Thrift, N. (1990). Class and change in rural Britain. In T. Marsden, P. Lowe and S. Whatmore (eds). *Rural restructuring: global processes and their responses*. Fulton, London.

Commission of the European Communities (CEC) (1985). *Perspectives for the Common Agricultural Policy*. CEC, Brussels.

Commission of the European Communities (CEC) (1987). *The agricultural situation in the Community: twentieth general report*. CEC, Brussels.

Commission of the European Communities (CEC) (1997). *Agenda 2000: for a stronger and wider Union*. CEC, Brussels.

Cooke, P. and Morgan, K. (1993). The network paradigm: new departures in corporate in corporate and regional development. *Environment and Planning C: Society and Space* 11, 543–564.

Cooke, P. and Morgan, K. (1998). *The associational economy*. Oxford University Press, Oxford.

Council for the Protection of Rural England (CPRE) (1995). *Memorandum to Select Committee on the Environment Second Report: Housing Need*. HMSO, London.

Council for the Protection of Rural England (CPRE) (1999). *Overview of Regional Development Agencies' Economic Strategies*. CPRE, London.

Council for the Protection of Rural England (CPRE) (2000). Press release – RDA's first year report: falling short on the environment. CPRE, London.

Council for the Protection of Rural England (CPRE) (2001) *Sprawl patrol: first year report*. CPRE, London.

Counsell, D. (1998). Sustainable development and structure plans in England and Wales: a review of current practice. *Journal of Environmental Planning and Management* 41, 177–194.

Counsell, D. (1999). Sustainable development and structure plans in England and Wales: operationalising the themes and principles. *Journal of Environmental Planning and Management* 42, 45–61.

Countryside Agency (2000). *The state of the countryside*. Countryside Agency, Cheltenham.

Countryside Agency (2001). *The state of the countryside*. Countryside Agency, Cheltenham.

Cox, G., Lowe, P. and Winter, M. (1986a). The state and the farmer: perspectives on agricultural policy. In G. Cox, P. Lowe and M. Winter (eds). *Agriculture: people and policies*. Allen and Unwin, London.

Cox, G., Lowe, P. and Winter, M. (1986b). From state direction to self regulation: the historical development of corporatism in British agriculture. *Policy and Politics* 14, 475–490.

Cox, G., Lowe, P. and Winter, M. (1987). Farmers and the state: a crisis for corporatism. *Political Quarterly* 54, 268–282.

Cross, D. (1990). *Counterurbanisation in England and Wales*. Avebury, Aldershot.

Cullingworth, B. (1997). British land use planning: a failure to cope with change? *Urban Studies* 5, 930–952.

Cullingworth, B. and Nadin, V. (1997). *Town and country planning in Britain* [12th edition]. Routledge, London.

Day, G. (1998). A community of communities? Similarity and difference in Welsh rural community studies. *The Economic and Social Review* 29, 233–257.

Day, G. and Fitton, M. (1975) Religion and social status in rural Wales. *The Sociological Review* 23, 867–891.

Day, G. and Murdoch, J. (1993). Locality and community: coming to terms with place. *The Sociological Review* 41, 82–111.

Day, G., Rees, G. and Murdoch, J. (1989). Social change, rural localities and the state: the restructuring of rural Wales. *Journal of Rural Studies* 5, 227–244.

Deakin, N. (2002). *In search of civil society*. Palgrave, London.

Department of the Environment (DoE) (1992). *Planning policy guidance note 12: development plans.* Department of the Environment, Transport and Regions (DETR), London.

Department of the Environment (DoE) (1996). *Household growth: where shall we live*. DoE, London.

Department of the Environment (DoE)/Ministry of Agriculture, Fisheries and Food (MAFF) (1995). *Rural England: a nation committed to a living countryside*. HMSO, London.

Department of the Environment (DoE)/Welsh Office (1992). *PPG 7: the countryside and the rural economy*. DoE, London.

Department of Environment, Food and Rural Affairs (DEFRA) (2002a). Statistical news release: provisional estimates of farm incomes in the UK, 2001. DEFRA, London.

Department of Environment, Food and Rural Affairs (DEFRA) (2002b). *Regional quality of life counts – 2001*. DEFRA, London.

Department of Environment, Food and Rural Affairs (DEFRA) (2002c). New voice for rural communities. *DEFRA Press Release 6/02*, 8 January. DEFRA, London.

Department of Environment, Transport and the Regions (DETR) (1998a). *The future of regional planning guidance*. DETR, London.

Department of Environment, Transport and the Regions (DETR) (1998b). *Planning for the communities of the future.* DETR, London.

Department of Environment, Transport and the Regions (DETR) (1998c) *Guidance to the Regional Development Agencies on rural policy*. DETR, London.

Department of Environment, Transport and the Regions (DETR) (1999). *Planning policy guidance note 11: regional planning: consultation draft*. DETR, London.

Department of Environment, Transport and the Regions (DETR) (2000a). *Planning policy guidance note 3: housing.* DETR, London.

Department of Environment, Transport and the Regions (DETR) (2000b). *Planning policy guidance note 11: regional planning*. DETR, London.

Department of Environment, Transport and the Regions (DETR) (2000c). *Draft regional planning guidance for the south east: proposed changes.* DETR, London.

Department of Environment, Transport and the Regions (DETR) (2000d). *Guidance on preparing Regional Sustainable Development Frameworks*. DETR, London.

Department of Environment, Transport and the Regions (DETR) (2001). *Regional planning guidance for the South East.* DETR, London.

Department of Environment, Transport and the Regions (DETR) and Ministry of Agriculture, Fisheries and Food (MAFF) (2000). *Our countryside: our future – a fair deal for rural England* [Cm 4909]. The Stationery Office, London.

Department of Transport, Local Government and the Regions (DTLR) (2001). *Planning green paper: planning: delivering fundamental change*. DTLR, London.

Department of Transport, Local Government and the Regions (DTLR) (2002). *Your region, your choice: revitalising the English regions.* DTLR, London.

Devon County Council (1998a) *Devon biodiversity action plan*. Devon County Council, Exeter.

Devon County Council (1998b). *Tourism trends in Devon 1997*. Devon County Council, Exeter.

Devon County Council (1998c). *Devon County Structure Plan first review 1995–2011: proposed modifications*. Devon County Council, Exeter.

Devon County Council (1999). *Working for a prosperous Devon*. Devon County Council, Exeter.

DTZ Pieda Consulting (1998). *Economic impact of BSE in the UK economy*. DTZ Pieda Consulting, Manchester.

DTZ Pieda Consulting (1999). *The South West of England: regional policy in a rural region*. Report to the South West of England Regional development Agency. DTZ Pieda, Reading.

Edwards, B., Goodwin, M., Pemberton, S. and Woods, M. (2001). Partnership, power and scale in rural governance. *Environment and Planning C: Government and Policy* 19, 289–310.

Emirbayer, M. and Goodwin, J. (1994). Network analysis, culture and the problem of agency. *American Sociological Review* 99, 1411–1454.

Fanfani, R. (1995). Agricultural change and agro-food districts in Italy. In D. Symes and A. Jansen (eds). *Agricultural restructuring and rural change in Europe*. Agricultural University, Wageningen.

Fennell, R. (1979). *The Common Agricultural Policy of the European Community*. Granada, St Albans.

Fennell, R. (1997). *The Common Agricultural Policy: continuity and change*. Clarendon, Oxford.

Fielding, A. (1990). Counterurbanisation: threat or blessing? In D. Pinder (ed.). *Western Europe: challenge and change*. Belhaven, London.

Fielding, A. (1992). Migration and social mobility: South East England as an escalator region. *Regional Studies* 26, 1–15.

Fielding, A. (1998). Counterurbanisation and social class. In P. Boyle and K. Halfacree (eds). *Migration into rural areas*. John Wiley, London.

Fothergill, S. and Gudgin, D. (1982) *Unequal growth*. Heinemann, London.

Gardner, B. (1996). *European agriculture: policies, production and trade*. Routledge, London.

Gilg, A. and Kelly, M. (1996). Farmers, planners and councillors: an insider's view of their interaction. In N. Curry and S. Owen (eds). *Changing rural policy in Britain*. Countryside and Community Press, Cheltenham.

Gillespie, A. (1999). The changing employment geography of Britain. In M. Breheny (ed.). *The people: where will they work?* TCPA, London.

Glyn-Jones, A. (1975). *Growing older in a South Devon town*. University of Exeter, Exeter.

Goodwin, M. (1998). The governance of rural areas: some emerging research issues and agendas. *Journal of Rural Studies* 14, 5–12.

Government Office for the South East (GOSE) (2000). *Housing technical note*. GOSE, Guildford.

Government Office for the South West (GOSW) (2001). *Regional planning guidance for the South West: consultation draft*. GOSW, Bristol.

Grant, W. (1973). Non-partisanship in British local politics. *Policy and Politics* 1, 241–254.

Grant, W. (1983). The NFU: the classic case of incorporation. In D. Marsh (ed.). *Pressure politics*. Junction Books, London.

Gripaios, R. (1994). The importance of the retirement industry. The South West economy – trends and prospects. Plymouth Business School, Plymouth.

Green, A. (1997). A question of compromise? Case study evidence on the location and mobility strategies of dual career households. *Regional Studies* 26, 44–59.

Halfacree, K. (1993) Locality and social representation: space, discourse and alternative definitions of the rural. *Journal of Rural Studies* 9, 23–37.

Halfacree, K. (1994). The importance of 'the rural' in the constitution of counterurbanisation: evidence from England in the 1980s. *Sociologia Ruralis* 34, 164–189.

Halfacree, K. (1995). Talking about rurality: social representations of the rural as expressed by residents of six English parishes. *Journal of Rural Studies* 11, 1–20.

Halfacree, K. (1996). Out of place in the country: travellers and the rural idyll. *Antipode* 28, 42–72.

Hall, P. (1998). Hot potatoes and bad omens. *Town and Country Planning*. July, 200.

Hall, P., Thomas, R., Gracey, H. and Drewett, R. (1973). *The containment of urban England.* Allen and Unwin, London.

Halliday, I. and Coombes, M. (1995). In search of counterurbanisation: some evidence from Devon on the relationship between patterns of migration and motivation. *Journal of Rural Studies* 11, 433–446.

Harper, S. (1989). The British rural community: an overview of perspectives. *Journal of Rural Studies* 5, 89–105.

Healey, P. (1999). Sites, jobs and portfolios: economic development discourses in the planning system. *Urban Studies* 36, 27–42.

Hetherington, K. (2000). *New Age travellers.* Cassells, London.

Hill, A. (2002). Acid house and Thatcherism: noise, the mob and the English countryside. *British Journal of Sociology* 53, 89–105.

Hinchcliffe, S. (2001). Indeterminacy in-decisions – science, science policy and politics in the BSE (bovine spongiform encephalopathy) crisis. *Transactions of the Institute of British Geographers* 26, 182–204.

Hodge, I. (2000). Countryside planning: from urban containment to sustainable development. In B. Cullingworth (ed.). *British planning: 50 years of urban and regional policy.* The Athlone Press, London.

Hoggart, K. (1990). Let's do away with the rural. *Journal of Rural Studies* 6, 245–257.

Hoggart, K. (1994). Review of Murdoch, J. and Marsden, T. 'Re-constituting Rurality'. *Journal of Rural Studies* 10, 90–92.

Hoggart, K. (1997). The middle classes in rural England, 1971–1991. *Journal of Rural Studies* 13, 253–273.

Hoggart, K. (1998). Rural cannot equal middle class because class does not exist? *Journal of Rural Studies* 14, 381–386.

Hoskins, W. G. (1959). *Devon and its people.* David and Charles, Newton Abbot.

House Builders Federation (HBF) (1998). *Memorandum to Select Committee on Environment, Transport and the Regions tenth report: housing.* The Stationery Office, London.

House of Commons (1998). *Select Committee on Environment, Transport and the Regions tenth report: housing.* HMSO, London.

Hudson, R. and Williams, A. (1995). *Divided Britain* [2nd edition]. John Wiley, London.

Ilbery, B. (1998). Dimensions of rural change. In B. Ilbery (ed.). *The geography of rural change.* Longman, London.

Ilbery, B. and Bowler, I. (1998). From agricultural productivism to post-productivism. In B. Ilbery (ed.). *The geography of rural change.* Longman, London.

John, P. and Whitehead, A. (1997). The renaissance of English regionalism in the 1990s. *Policy and Politics* 25, 7–17.

Jonsson, C., Tagil, S. and Tornqvist, G. (2000). *Organising European space.* Sage, London.

Kay, A. (1998) *The reform of the Common Agricultural Policy: the case of the McSharry reforms*. CAB International, Wallingford.

Keating, M. (1997). The innovation of regions: political restructuring and territorial government in Western Europe. *Environment and Planning C: Government and Policy* 15, 383–398.

Keeble, D. and Nachum, L. (2002). Why do business service firms cluster? Small consultancies, clustering and decentralisation in London and southern England. *Transactions of the Institute of British Geographers* NS 27, 67–90.

Keeble, D. and Tyler, P. (1995). Enterprising behaviour and the urban–rural shift. *Urban Studies* 32, 975–997.

Keeble, D., Tyler, P., Broom, G. and Lewis, J. (1992). *Business success in the countryside: the performance of rural enterprise*. HMSO, London.

King, D. (2000). *Projected household numbers for rural districts of England*. Countryside Agency, Chelmsford.

Kirk, J. (1979). *The development of agriculture in Germany and the UK: UK agricultural policy 1870–1970*. Wye College, Ashford.

Kneale, J., Lowe, P. and Marsden, T. (1992). The conversion of agricultural buildings: an analysis of variable pressures and regulations towards the post-productivist countryside. ESRC Countryside Change Initiative Working Paper Series 29. University of Newcastle, Newcastle upon Tyne.

Knoke, D. and Kuklinski, J. (1982). *Network analysis*. Sage, London.

Lamont, M. and Thevenot, L. (2000). Introduction: toward a renewed comparative cultural sociology. In M. Lamont and L. Thevenot (eds). *Rethinking comparative cultural sociology: repertoires of evaluation in France and the United States*. Cambridge University Press, London.

Laschewski, L., Phillipson, J. and Gorton, M. (2002) The facilitation and formalisation of small business networks: evidence from the North East of England. *Environment and Planning C: Government and Policy* 20, 375–391.

Lash, S. and Urry, J. (1994). *Economies of signs and space*. Sage, London.

Latour, B. (1987). *Science in action*. Open University Press, Milton Keynes.

Latour, B. (1993). *We have never been modern*. Harvester Wheasheaf, Hemel Hempstead.

Lawrence, M. (1997). Heartlands or neglected geographies? Liminality, power and the hyperreal rural. *Journal of Rural Studies* 13, 1–17.

Le Gales, P. and Voelzkow, H. (2001). The governance of local economies. In C. Crouch, P. le Gales, C. Triglia and H. Voelzkow (eds). *Local production systems in Europe: rise or demise?* Oxford University Press, Oxford.

Lewis, G. (1998). Rural migration and demographic change. In B. Ilbery (ed.). *The geography of rural change*. Longman, London.

Lewis, G. (2000). Changing places in a rural world: the population turnaround in perspective. *Geography* 85, 157–165.

Liepins, R. (2000). New energies for an old idea: reworking approaches to 'community' in contemporary rural studies. *Journal of Rural Studies* 16, 23–35.

Little, A. (2002). *The politics of community*. Edinburgh University Press, Edinburgh.

Lobley, M. and Potter, C. (1998). Environmental stewardship in UK agriculture: a comparison of the Environmentally Sensitive Areas Programme and the Countryside Stewardship Scheme in South East England. *Geoforum* 29, 413–432.

Lock, D. (2000). Degrees of purgatory. *Town and Country Planning* 69, 126–130.

Lowe, P. (1997). The British Rural White Papers: a comparison and critique. *Journal of Environmental Planning and Management* 40, 389–400.

Lowe, P. and Buller, H. (1990). Overview. In P. Lowe and M. Bodiguel (eds). *Rural studies in Britain and France*. Belhaven, London.

Lowe, P., Clark, J., Seymour, S. and Ward, N. (1997). *Moralising the environment*. UCL Press, London.

Lowe, P., Cox, G., O'Riordan, T., MacEwan, M. and Winter, M. (1986). *Countryside conflicts: the politics of farming, forestry and conservation.* Gower, Aldershot.

Lowe, P., Murdoch, J. and Cox, G. (1995). A civilised retreat? Anti-urbanism, rurality and the making of an Anglo-centric culture. In P. Healey *et al.* (eds) *Managing cities: the new urban context.* John Wiley, London.

Lowe, P., Murdoch, J., Marsden, T., Munton, R. and Flynn, A. (1993). Regulating the new rural spaces: the uneven development of land. *Journal of Rural Studies* 9, 205–222.

Lowe, P., Murdoch, J. and Norton, A. (2001). *Professionals and volunteers in the environmental process*. Centre for Rural Economy, University of Newcastle, Newcastle.

Lowe, P., Ray, C., Ward, N., Wood, D. and Woodward, R. (1998). *Participation in rural development: a review of European experience.* Centre for Rural Economy Research Report, Department of Agricultural Economics and Food Marketing, University of Newcastle, Newcastle upon Tyne.

Lowe, P. and Ward, N. (1997) Field-level bureaucrats and the making of new moral discourses in agri-environmental controversies. In D. Goodman and M. Watts (eds). *Globalising food: agrarian questions and global restructuring*. Routledge, London.

Lowe, P. and Ward, N. (1998). Regional policy, CAP reform and rural development in Britain: the challenge for New Labour. *Regional Studies* 32, 469–474.

Lowenthal, D. (1985). *The past is a foreign country.* Cambridge University Press, Cambridge.

MacLeod, G. (2001). New regionalism reconsidered: globalisation and the remaking of political economic space. *International Journal of Urban and Regional Research* 4, 804–829.

MacLeod, G. and Goodwin, M. (1999). Space, scale and state strategy: rethinking urban and regional governance. *Progress in Human Geography* 23, 503–527.

MacLeod, G. and Jones, M. (2001). Re*newing* the geography of regions. *Environment and Planning D: Society and Space* 19, 669–695.

McCord, N. (1979). *North East of England*. Batsford Academic, London.

Maillat, D. (1996). Regional productive systems and innovative milieux. In OECD (ed.). *Networks of enterprises and local development.* OECD, Paris.

Malmberg, A. and Maskell, P. (2002). The elusive concept of localisation economies: towards a knowledge-based theory of spatial clustering. *Environment and Planning A* 34, 429–449.

Marsden, T. (1992). Exploring a rural sociology for the Fordist transition: incorporating social relations into economic restructuring. *Sociologia Ruralis* 32, 209–230.

Marsden, T. (1998). New rural territories: regulating the differentiated rural spaces. *Journal of Rural Studies* 14, 107–117.

Marsden, T., Murdoch, J., Lowe, P., Munton, R. and Flynn, A. (1993). *Constructing the countryside.* UCL Press, London.

Marsden, T., Munton, R., Ward, N. and Whatmore, S. (1996). Agricultural geography and the political economy approach. *Economic Geography* 72, 361–376.

Marsh, J. (1982). *Back to the land: the pastoral impluse in England, from 1880 to 1914.* Quartet, London.

Massey, D. (1984). *Spatial divisions of labour.* Macmillan, London.

Massey, D. (1991). A global sense of place. *Marxism Today* June, 24–29.

Matless, D. (1998). *Landscape and Englishness*. Reaktion, London.

Mawson, J. and Spencer, K. (1997). The Government Offices for the English Regions: towards regional governance? *Policy and Politics* 25, 71–84.

Milbourne, P. (ed.) (1997). *Revealing rural 'others': representation, power and identity in the British countryside*. Pinter, London.

Ministry of Agriculture, Fisheries and Food (MAFF) (1975). *Food from our own resources*. MAFF, London.

Ministry of Agriculture, Fisheries and Food (MAFF) (1979). *Farming and the nation*. MAFF, London.

Ministry of Agriculture, Fisheries and Food (MAFF) (1999). *Reducing farm subsidies – economic adjustment in rural areas*. MAFF, London.

Monk, S. and Hodge, I. (1995). Labour markets and employment opportunities in rural Britain. *Sociologia Ruralis* 35, 153–172.

Mormont, M. (1990). Who is rural? Or, how to be rural: towards a sociology of the rural. In T. Marsden, P. Lowe and S. Whatmore (eds). *Rural restructuring: global processes and their local response*. Fulton, London.

Mosely, M. (2000). Innovation and rural development: some lessons from Britain and Western Europe. *Planning, Practice and Research* 15, 95–115.

Munton, R. (1983). *London's green belt: containment in practice*. Allen and Unwin, London.

Munton, R. (1995). Regulating rural change: property rights, economy and environment – a case study from Cumbria, UK. *Journal of Rural Studies* 11, 269–284.

Murdoch, J. (1988). *State and agriculture in Wales*. Unpublished PhD thesis, University of Wales, Aberystwyth.

Murdoch, J. (1992). Representing the region: Welsh farmers and the British state. In T. Marsden, P. Lowe and S. Whatmore (eds). *Labour and locality: Vol. IV* of *Critical perspectives on rural change*. Fulton, London.

Murdoch, J. (1995a). Governmentality and the politics of resistance in UK agriculture: the case of the Farmers' Union of Wales. *Sociologia Ruralis* 35, 187–205.

Murdoch, J. (1995b) Middle-class territory? Some remarks on the use of class analysis in rural studies. *Environment and Planning A* 27, 1213–1230.

Murdoch, J. (1996). Planning the rural economy. In P. Allanson and M. Whitby (eds). *The rural economy and the British countryside*. Earthscan, London.

Murdoch, J. (1997a). Towards a geography of heterogeneous associations. *Progress in Human Geography* 21, 321–337.

Murdoch, J. (1997b). The shifting territory of government: some insights from the Rural White Paper. *Area* 29, 109–118.

Murdoch, J. (1998). Counterurbanisation and the countryside: some causes and consequences. *Papers in Environmental Planning Research* 15. Department of City and Regional Planning, University of Wales, Cardiff.

Murdoch, J. (2000). Networks – a new paradigm of rural development? *Journal of Rural Studies* 16, 407–419.

Murdoch, J. and Abram, S. (1998). Defining the limits of community governance. *Journal of Rural Studies* 14, 41–50.

Murdoch, J. and Abram, S. (2002). *Rationalities of planning: development versus environment in planning for housing*. Ashgate, Aldershot.

Murdoch, J. Abram, S. and Marsden, T. (1999). Modalities of planning: a reflection on the persuasive powers of the development plan. *Town Planning Review* 70, 191–212.

Murdoch, J., Abram, S. and Marsden, T. (2000). Technical expertise and public participation in planning for housing: 'playing the numbers game'. In G. Stoker (ed.). *Power and participation: the new politics of local governance*. Macmillan, London.

Murdoch, J. and Day, G. (1998). Middle-class mobility, rural communities and the politics of exclusion. In P. Boyle and K. Halfacree (eds). *Migration into rural areas: theories and issues*. John Wiley, London.

Murdoch, J. and Marsden, T. (1994). *Reconstituting rurality: class, community and power in the development process*. UCL Press, London.

Murdoch, J. and Marsden, T. (1995). The spatialisation of politics: local and national actor-spaces in environmental conflict. *Transactions of the Institute of British Geographers* NS 20, 368–380.

Murdoch, J. and Pratt, A. C. (1993). Rural studies: modernism, post-modernism and the post-rural. *Journal of Rural Studies* 9, 411–427.

Murdoch, J. and Pratt, A. C. (1994) Rural studies of power and the power of rural studies: a reply to Philo. *Journal of Rural Studies* 10, 83–87.

Murdoch, J. and Ward, N. (1997). Governmentality and territoriality: the statistical manufacture of Britain's 'national farm'. *Political Geography* 16, 307–324.

Murray, K. (1955). *Agriculture: a history of the Second World War*. HMSO, London.

Newby, H. (1977). *The deferential worker*. Allen Lane, London.

Newby, H. (1979). *Green and pleasant land*. Penguin, London.

Newby, H. (1981). Rural sociology: a trend report. *Current Sociology* 12, 1–132.

Newby, H. (1985). *Green and pleasant land* [2nd edition]. Penguin, Harmondsworth.

Newby, H. (1987). *Country life*. Weidenfield and Nicholson, London.

Newby, H., Bell, C., Rose, D. and Saunders, P. (1978). *Property, paternalism and power*. Hutchinson, London.

North, D. (1998). Rural industrialisation. In B. Ilbery (ed.). *The geography of rural change*. Longman, London.

North, D. and Smallbone, D. (2000). The innovativeness and growth of rural SMEs during the 1990s. *Regional Studies* 34, 145–157.

Northern Infomatics (1997). *Avoiding exclusion: the challenge of shaping the information society in the rural north*. Northern Infomatics, Sunderland.

One North East (2002). *Realising our potential: consultation draft*. One North East, Newcastle upon Tyne.

Paasi, A. (1991). Deconstructing regions: notes on the scales of spatial life. *Environment and Planning A* 23, 239–256.

Pahl, R. (1966). *Urbs in rure*. London School of Economics, London.

Pahl, R. (1970). *Readings in urban sociology*. Pergamon, Oxford.

Paxman, J. (1991). *Friends in high places: who runs Britain?* Penguin, London.

Performance and Innovation Unit (PIU) (1999). *Rural economies*. Cabinet Office, London.

Phillips, M. (1998). Investigation of the British rural middle classess – part 1: from legislation to interpretation. *Journal of Rural Studies* 14, 411–425.

Philo, C. (1992). Neglected rural geographies: a review. *Journal of Rural Studies* 8, 193–207.

Policy Commission on the Future of Food and Farming (2002). *Farming and food: a sustainable future*. Cabinet Office, London.

Potter, C. (1998). *Against the grain: agri-environmental reform in the United States and the European Union*. CAB International, Wallingford.

Powell, W. and Smith-Doerr, L. (1994). Networks and economic life. In N. Smelser and R. Swedberg (eds). *The handbook of economic sociology*. Sage, London.

Pratt, A. (1996). Discourses of rurality: loose talk or social struggle? *Journal of Rural Studies* 12, 69–78.

Public and Corporate Consultants (2000). *The economic effects of hunting with dogs: research report to the Home Office Committee of Inquiry into Hunting with Dogs*. The Stationery Office, London.

Raley, M. and Moxey, A. (2000). *Rural micro-businesses in North East England*. Centre for Rural Economy Research Report, University of Newcastle, Newcastle upon Tyne.

Rapport, N. (1994). *Diverse world views in an English village*. Edinburgh University Press, Edinburgh.

Ray, C. (1998). Territory, structures and interpretation – two case studies of the European Union's LEADER 1 programme. *Journal of Rural Studies* 14, 79–88.

Reade, E. (1987). *The British town and country planning system.* Open University Press, Milton Keynes.

Rees, G. (1984). Rural regions in national and international economies. In T. Bradley and P. Lowe (eds). *Locality and rurality.* Geo Books, Norwich.

Regional Trends (1998). *Regional trends.* The Stationary Office, London.

Report of Panel (1999). *Regional planning guidance for South East of England.* Government Office for the South East, Guildford.

Rhodes, R. (1997). *Understanding governance: policy networks, governance, reflexivity and accountability.* Open University Press, Buckingham.

Rogers, A. (1993). *English rural communities: an assessment and prospects for the 1990s.* Rural Development Commission, London.

Rogers, A. (1999). *The most revolutionary measure: a history of the Rural Development Commission 1909–1999.* Rural Development Commission, London.

Rydin, Y. (1999). Public participation in planning. In B. Cullingworth (ed.). *British planning: 50 years of urban and regional policy.* The Athlone Press, London.

Saraceno, E. (1993). Alternative readings of spatial differentiation: the rural versus the local economy approach in Italy. *European Review of Agricultural Economics* 21, 451–474.

Savage, M., Barlow, J., Dickens, P. and Fielding, T. (1992). *Property, bureaucracy and culture: middle-class formation in contemporary Britain.* Routledge, London.

Savage, M., Stovel, K. and Bearman, P. (2001). Class formation and localism in an emerging bureaucracy: British bank workers, 1880–1960. *International Journal of Urban and Regional Research* 25, 284–305.

Scottish Office (1995). *Rural Scotland: people, prosperity and partnership*. Scottish Office, Edinburgh.

Self, P. and Storing, P. (1962). *The state and the farmer.* Allen and Unwin, London.

Shaw, R. (1997). The Rural White Paper in England: the origins, production and immediate consequences of the White Paper 'Rural England'. *Journal of Environmental Planning and Management* 40, 381–385.

Sheail, J. (1997). Scott revisited: post-war agriculture, planning and the countryside. *Journal of Rural Studies* 13, 387–398.

Shoard, M. (1979). *The theft of the countryside*. Temple Smith, London.

Short, J. R. (2001). *Global dimensions: space, place and the contemporary world*. Reaktion, London.

Short, J. R., Fleming, S. and Witt, S. (1986). *House building, planning and community action.* Routledge and Kegan Paul, London.

Short, B., Watkins, C., Foot, W. and Kinsman, P. (2000). *The National Farm Survey 1941–1943: state surveillance and the countryside in England and Wales in the Second World War*. CAB International, Wallingford.

Silk, J. (1999). The dynamics of community, place and identity. *Environment and Planning A* 31, 5–17.

Smith, A. (2000). Policy networks and advocacy coalitions: explaining policy change and stability in UK industrial pollution policy. *Government and Policy* 18, 95–114.

Smith, M. (1989). The Annual Review: the emergence of a corporatist institution? *Political Studies* 37, 81–96.

South East Regional Planning Forum (SERPLAN) (1997). South East Regional Planning Capability Study: an introductory note. SERPLAN, London.

South East Regional Planning Forum (SERPLAN) (1998). *A sustainable development strategy for the South East: consultation document*. SERPLAN, London.

South East Regional Planning Forum (SERPLAN) (1999). *Regional planning guidance for the South East: report of panel: SERPLAN's response*. SERPLAN, London.

Staber, U. (2001). The structure of networks in industrial districts. *International Journal of Urban and Regional Research* 25, 537–552.

Stacey, M. (1969). The myth of community studies. *British Journal of Sociology*. 20, 34–47.

Stanes, R. (1985). *A history of Devon*. Phillimore, Exeter.

Stanyer, J. (1978) *Farming politics: Devon and Cheshire compared*. Open University D203, Block III. Part 8, 275–285.

Stanyer, J. (1989) *A History of Devon County Council*. Devon Books, Exeter.

Storper, M. (1997). *The regional world*. The Guilford Press, London.

Storper, M. and Salais, R. (1997). *Worlds of production*. Harvard University Press, Cambridge, MA.

Talbot, H. (1997). *Rural telematics in England: strategic issues*. Centre for Rural Economy Research Report, Department of Agricultural Economics and Food Marketing, University of Newcastle, Newcastle upon Tyne.

Tarling, R., Rhodes, J., North, J. and Broom, G. (1993). *The economy of rural England*. Rural Development Commission, London.

Thevenot, L. (2002). Which road to follow? The moral complexity of an 'equipped' humanity. In J. Law and A. M. Mol (eds). *Complexities: social studies of knowledge practices*. Duke University Press, Durham, NC.

Thevenot, L., Moody, M. and Lafaye, C. (2000). Forms of valuing nature: arguments and modes of justification in French and American environmental disputes. In M. Lamont and L. Thevenot (eds). *Rethinking comparative cultural sociology: repertoires of evaluation in France and the United States*. Cambridge University Press, Cambridge.

Thrift, N. (1987). Manufacturing rural geography. *Journal of Rural Studies* 3, 77–81.

Thrift, N. (1989). Images of social change. In C. Hamnett, L. McDowell and P. Sarre (eds). *The changing social structure*. Sage, London.

Tracey, M. (1982). *Agriculture in Western Europe: challenge and response 1880–1980*. Granada, St Albans.

Travis, J. F. (1993). *The rise of Devon seaside resorts 1750–1900*. University of Exeter Press, Exeter.

Turok, I. and Edge, N. (1999). *The jobs gap in Britain's cities: employment loss and labour market consequences*. Policy Press, Bristol.

Urban Task Force (1999). *Towards urban renaissance*. DETR, London.

Urry, J. (1984). Capitalist restructuring, recomposition and the regions. In T. Bradley and P. Lowe (eds). *Locality and rurality*. Geo Books, Norwich.

Urry, J. (1995). *Consuming places*. Routledge, London.

Urry, J. (2000). *Sociology beyond societies: mobilities for the twenty-first century*. Routledge, London.

Valentine, G. (2000). *Social geographies*. Prentice-Hall, Harlow.

van der Ploeg, J. D. and van Dijk, G. (eds) (1995). *Beyond modernisation: the impact of endogenous rural development*. Van Gorcum, Assen, the Netherlands.

Vigar, G., Healey, P., Hull, A. and Davoudi, S. (2000). *Planning, governance and spatial strategy in Britain: an institutionalist approach*. Macmillan, London.

Walker, R. (1997). Field of dreams or the best game in town? In D. Goodman and M. Watts (eds). *Globalising food: agrarian questions and global restructuring*. Routledge, London.

Wannop, U. and Cherry, G. (1994). The development of regional planning in the UK. *Planning Perspectives* 9, 29–60.

Ward, N. (1999). Foxing the nation: the economic (in)significance of hunting with hounds in Britain. *Journal of Rural Studies* 15, 389–403.

Ward, N., Lowe, P., Bridges, T., Stafford, J. and Evans, N. (2001). *Regional Development Agencies and rural development: priorities for action*. East of England Development Agency.

Ward, N. and McNicholas, K. (1998). Reconfiguring rural development in the UK: Objective 5b and the new rural governence. *Journal of Rural Studies* 14, 27–40.

Ward, S. (1994). *Planning and urban change*. Paul Chapman Publishing, London.

Warde, A. (1985). The homogenisation of space? Trends in the spatial division of labour. In H. Newby, J. Bujra, P. Littlewood, G. Rees and T. Rees (eds) (1985). *Restructuring capital: recession and reorganisation in industrial society*. Macmillan, London.

Welsh Office (1996). *A working countryside for Wales*. Welsh Office, Cardiff.

West Country Tourist Board (1992). *Devon tourism holiday-taking behaviour research*. West Country Tourist Board, Exeter.

Whitby, M. (ed.) (1996). *The European environment and CAP reform: policies and prospects for conservation*. CAB International, Wallingford.

Whitby, M., Townsend, A., Gorton, M. and Parsisson, D. (1999). *The rural economy of North East England*. Centre for Rural Economy Research Report, University of Newcastle, Newcastle upon Tyne.

Williams, R. (1973). *The country and the city*. Chatto and Windus, London.

Williams, W. (1963). *A West Country village: Ashworthy*. Routledge and Kegan Paul, London.

Wilson, G. (2001). From productivism to post-productivism . . . and back again? Exploring the (un)changed natural and mental landscapes of European agriculture. *Transactions of the Institute of British Geographers* NS 26, 77–102.

Winnifrith, J. (1962). *The Ministry of Agriculture, Fisheries and Food*. Allen and Unwin, London.

Winter, D. M. (1986). *The survival and re-emergence of family farming: a study of the Holsworthy area of West Devon*. Unpublished PhD thesis, Open University, Milton Keynes.

Winter, M. (1996). *Rural politics*. Routledge, London.

Wittel, A. (2001). Towards a network sociality. *Theory, Culture and Society* 18, 51–76.

Woodward, R. (1998). Gunning for England: the politics of the promotion of military land use in the Northumberland National Park. *Journal of Rural Studies* 15, 17–33.

Wormell, P. (1978). *The anatomy of agriculture: a study of Britain's greatest industry*. Harper and Row, London.

Index

Note: 'n.' after a page reference indicates the number of a note on that page.